D0408968

PATTERN BIOLOGY
and the Complex Architectures of Life

PATTERN BIOLOGY and the Complex Architectures of Life

Michael J. Katz

Longwood Academic
Wolfeboro, NH

Published in 1988 by
LONGWOOD ACADEMIC
A Division of Longwood Publishing Group, Inc.
27 South Main Street
Wolfeboro, NH 03894-2069
USA

ISBN 0-89341-521-9

Library of Congress Cataloging in Publication Data:

Katz, Michael J.
Pattern Biology and the complex architectures of life.

Companion volume to: Templets and the explanation of complex patterns.
Bibliography: p.
Includes index.
1. Biochemical templates. 2. Morphology—Philosophy.
3. Biology—Philosophy. 4. Life (Biology) I. Katz,
Michael J. Templets and the explanation of complex patterns. II. Title.
QP517.B48K37 1987 574 87-22651
ISBN 0-89341-521-9

Printed in the United States of America

To
my children

ETHAN AT AGE 6½

Color-man with caftan
Castle-webbed scrapman
Tumble-jum jongleur

Forest-cribbed creatures
Climbing from my land
Into his small hands.

and

EMILY ANNE AT AGE 1¾

Stumpy from a distance squat
  stub of an Emily Anne
Collywobbly spider caught
  inside her tight little hand

She jumps and runs through grassy lawns
  edged with thick green vines
And pale skies roll around her dawns
  in somersault baby designs

A tiny gold-haired child who ought
  still hold my thumb wide-eyed
Till the clouds will stop to watch the plots
  of men risen again who'd died.

CONTENTS

ACKNOWLEDGEMENTS

Pattern Biology is the companion to my *Templets and the Explanation of Complex Patterns*, Cambridge: Cambridge University Press, 1986. Both volumes attempt to block out vocabulary, concepts, questions, and methodologies that can be useful to those scientists who see all the world as patterns, patterns, and more patterns.

It was one sunny February morning in 1979 when Raymond J. Lasek proposed the name "Pattern Biology." Names are important—they cast lineaments upon our amorphous ideas—and I believe Lasek had named quite a fine name. Thanks, Ray.

Frederick Sanger kindly wrote a brief reminiscence of how he first chose to work on the insulin molecule.

The Alfred P. Sloan Foundation and the Whitehall Foundation provided generous support as I worked on my various pattern projects.

"'Tis the Gift to be Simple" first appeared in the *Kenyon Review*, vol. 9 (1987). An early version of "The Luxuriance of Nature" first appeared as "Are There Biological Impossibilities?" in *No Way: The Nature of the Impossible*, eds. P.J. Davis and D. Park, New York: W.H. Freeman, 1987.

The author is grateful for permission for quotations from the following: "The Creation," *God's Trombones* by James Weldon Johnson; copyright 1927 by The Viking Press; copyright renewed 1955 by Grace Nail Johnson; all rights reserved; reprinted by permission of Viking Penguin Inc. "How to Tell Corn Fairies if You See 'Em," from *Rootabaga Stories* by Carl Sandburg; copyright 1922, 1923 by Harcourt Brace Jovanovich, Inc.; renewed 1950, 1951 by Carl Sandburg; reprinted by permission of Harcourt Brace Jovanovich, Inc. "what if a much of a which of a wind" and "neither could say" from *Complete Poems 1913-1962* by e.e. cummings; copyright 1944 by e.e. cummings; renewed 1972 by Nancy T. Andrews; reprinted by permission of Harcourt Brace Jovanovich. "Ars Poetica," from *New and Collected Poems 1917-1976* by Archibald MacLeish; copyright 1976 by Archibald MacLeish; reprinted by permission of Houghton Mifflin Company. "The Maze," from *W.H. Auden: Collected Poems*, edited by Edward Mendelson; copyright 1966 by W.H. Auden; reprinted by permission of Random House Inc. "Looking for a sunset bird in winter," from *Collected Poems by Robert Frost*; reprinted by permission of Holt, Rinehart & Winston. "Spring morning," from *When We Were Very Young* by A.A. Milne; reprinted by permission of E.P. Dutton.

Special thanks to Mr. Archie Lieberman for permission to use his photograph "Aunt Hattie and Baby Melissa" on the cover of the book; copyright 1986 by Land's End, Inc.

M.J. Katz

INTRODUCTION

I

Patterns—they are the orderly assemblages that we attend to in our lives. When we are startled awake at midnight, the world suddenly washes over us in a tidal blur; when we are fevered, the room presses down like a thick cotton cloud; when we are stunned by a catastrophe, our surroundings recede into a distant vague movie; but always we snap back, sensations are sorted out and sensible patterns reappear. The mind is an automatic pattern generator: we are compelled to build orderly architectures from our sensory impressions, and these orderly architectures are our world patterns.

All patterns have two aspects. First, there are the constituent unit elements—a pattern's nuts and bolts, its content. In addition, there are the interactions and the arrangements among the elements—a pattern's struts and scaffolding, its configuration. Together content and configuration compose a pattern, and the meaning of a pattern stems as much from its configuration as from its constituent units. Technically, the information contained in a pattern is measured by how many different configurations could have been laid upon its same content: the most informative patterns come from large families with identical content and diverse configurations. And, as more information is packed into a pattern, more details of its configuration must be explicitly specified in order to capture that information.

Recreating patterns in a common, thoroughly-understood, and abstract formalism can set loose the deepest power of our abstractionist toolbox—this power is mathematics. With the partitioning of patterns into content and configuration comes a natural abstract representation, the adjacency matrix. The rows of an adjacency matrix summarize the content, and the slots summarize the configuration. For example, T-H-U-M-B is a simple linear pattern, and in matrix form it might be abstracted as:

		T	H	U	M	B
	T	0	1	0	0	0
	H	1	0	1	0	0
A=	U	0	1	0	1	0
	M	0	0	1	0	1
	B	0	0	0	1	0

The particular interrelation that is captured in this matrix is spatial adjacency. (The first row is read as follows: the unit element "T" is not spatially adjacent to itself, it is adjacent to "H", and is not adjacent to "U", "M", or "B".) Such matrices are well-studied structures with ties to most branches of mathematics,* and, when I would like to deal with patterns precisely, I will talk in matrices.

---

*For example, one useful mathematical characterization of a matrix is its *dimension*, the number of rows (in a square matrix). An nxn matrix *A* has *n* rows and n^2 slots; it has dimension $D_{max}=n$ and size $S_{max}=n^2$. D_{max} is the maximum potential dimensionality of *A*—it is the dimensionality in a universe where all unit elements are ideal and can potentially interact. D_{max} and S_{max} are a priori and minimal-constraint assumptions; however, in any real world situation the actual unit elements may be more limited in their inherent interactive capabilities. The actual available interactions (as dictated by the inherent natures of the unit elements) can be summarized in another matrix *M*. *M* is a binary matching matrix, and the sum of the rows of *M* is *S*, the effective size of *A*. [Likewise, the effective dimensionality of *A* is D=sqr(S).] $Q=S/S_{max}$, where *Q* is the templeting index of *A*. When the effective dimensionality of *A* is reduced by the inherent interactive potentials of the unit elements, then $Q \rightarrow O$, *A* is on the self-assembling end of the spectrum and the fabrication of *A* from its natural precursors is relatively determinate. Self-assembly breeds determinacy.

Determinate pattern-assembly can also be ensured in another manner. Patterns are always generated as the products of two classes of predecessors—unit elements and templets. When the inherent properties of the unit elements do not constrain them to a few final configurations, then $Q \rightarrow 1$ and the pattern-assembly system is extensively templeted. Here, determinancy can arise when the templets themselves are faithfully stereotyped. Faithfully stereotyped templets are self-assembling templets—templets constructed of elements that can be interlocked only in a small number of different configurations. (For example, an infrangible and long-lived templet is self-assembling, because every time it appears its unit elements are to be found only in one configuration.) The special patterns of biology are extensively templeted and determinate, and their determinacy emanates from their self-assembling templets.

II

The biologist is immersed in a thick web of patterns, especially the complex patterns, and these patterns are often poised between determinacy and stochasticism. Organic patterns have a bit of magic, they are at once idiosyncratic and unique yet recurrent and stereotyped. As Loren Eiseley [p.209], tramping out into the cold autumn fields and examining the bits and pieces of last summer's rush of life, recounted:

> Beautiful, angular, and bare the machinery of life will lie exposed...to my view. There will be the thin, blue skeleton of a hare tumbled in a little heap, and crouching over it I will marvel...at the wonderful correlation of parts, the perfect adaptation to purpose, the individually vanished and yet persisting pattern which is now hopping on some other hill. I will wonder, as always, in what manner "particles" pursue such devious plans and symmetries. I will ask once more in what way it is managed, that the simple dust takes on a history and begins to weave these unique...apparitions in the stream of time. I shall wonder what strange forces at the heart of matter regulate the tiny beating of a rabbit's heart or the dim dream that builds a milkweed pod.

Against the manifold landscape of the other natural sciences, biology is most characterized by those natural patterns that are at once complex and determinate.

In terms of patterns, the practical operations of a biologist can be classified as definition, explanation, and chronicling. Definition discerns the architecture of a pattern, explanation describes the formation of a pattern, and chronicling documents the consequences of a pattern. Different patterns are the natural and indigenous denizens of the different biological specialties: the natural patterns are natural within the particular frame of reference defined by each realm. But, although the constituent units differ from realm to realm, biological patterns can be equally complex in psychology, in cell biology, and in biochemistry, and the operations of pattern analysis remain the same throughout all fields.

III

Patterns range from simple to complex; simplicity and complexity identify the two ends of a spectrum. Simple things are small and uniform, complex things are large and diversely multipartite. Simple is homogeneous; complex is heterogeneous—and complexity complicates the pattern biologist's theories.

Just as patterns can be arrayed on the spectrum of simple to complex, so can the explanations of patterns. A simple explanation translates its phenomenon into units that are smooth and pocket-size, and simple things can be fully explained by such compact explanations. In contrast, truly complex phenomena cannot be effectively reduced in this fashion—they cannot be fully translated into a shorter, briefer, or less intricate set of components. Complex things are made of many varied building blocks, and the building blocks are interconnected in intricate ways. While simple things are like a tiny row of minnows, complex things are like a tangle of octopus or squid—some arms are interwoven and knotted together and some arms are waving free, and agile as you may be you cannot wrestle the whole cephalopodic horde into a bag without changing its protean architecture. For a complete understanding, complex and tangled phenomena cannot be neatly compacted: complex things require complex explanations.

The natural world is complex, and it has pockets of incondensable complexity, which appear most clearly in the realm of biology. This frustrates the biologist, because scientifically we aspire to complete explanations. (Leonardo da Vinci wrote that "abbreviators of works do harm to knowledge...for certainty springs from a complete knowledge of all the parts which united compose the whole." [Keele, p. 4]) To fully explain incondensable complexity, we need complex explanations—at times they cannot be avoided. At the same time, there is a countervailing force: humans comprehend best through simplicity. This poses the essential dilemma:

> Every generalisation supposes in a certain measure a belief in the unity and simplicity of Nature. [However,] it is not certain that Nature is [actually] simple...[The belief in simplicity is a] custom imposed upon physicists by [the need to have small and manageable theories], but how can it be

justified in the presence of discoveries which daily show us fresh details, richer and more complex?... We must stop somewhere, and for science to be possible we must stop where we have found simplicity. That is the only ground on which we can erect the edifice of our generalisations.

[Poincare, pp. 145-146 and 149]

—so wrote Henri Poincare, mathematician and physicist.

The biologist, however, does not always have the luxury of stopping with simplicity: when biologists strive for full explanations, they can be inexorably driven to complex explanations. Nature is sometimes complex; simplicity is not always the natural truth—Hans Reichenbach wrote:

One... system can be simpler than another; but that fact does not make it "truer" than the other. The decimal system is simpler than the yard-foot-inch system; but an architect's plan drawn in feet and inches is as true a description of a house as a plan drawn in the decimal system. A simplicity of this kind...is not a criterion of truth.

[Reichenbach, p. 296]

In reality, Nature uses "yard-foot-inch" system as comfortably as she uses "decimal" systems, because she is not an engineer and she need not proceed parsimoniously. Instead, Nature meanders along parallel paths, building overdetermined systems and luxuriating in convolutions and in excesses, and "[natural principles] are enveloped by so many strange circumstances that it requires great sagacity to disentangle them." [Laplace, pp. 176-177]

We can see Nature's Byzantine workings, for example, in teeth. Not long ago my first child was five and a half years old; by that time he had a great variety of experiences, but the inner clockworks of the world remained arcane. (He explained to me for instance that winter comes because it is cloudy—in December the clouds block off the sun completely, the sky gets chilly, and drops of water come down as snow.) In those days, he was worried by shots and also by cavities. He never had a cavity, and he wanted to know why we get holes in our teeth and will they go away and do they hurt. There was a spell when cavities came up quite often—perhaps his kindergarten class was learning how to take care of their teeth, or perhaps he was attending to toothpaste commercials on television—in any case, I came home one day and learned that he had a cavity.

"How do you know?" I asked, and he promptly showed me a brown spot on the front of one of his teeth. I suggested that it was just a bit of lunch, but toothbrushing left it unchanged, and my wife thought that it was time to see a dentist. I reverted to my more magical heritage, offering that the human body would cure itself; inertia prevailed, and we did nothing. In spite of my best efforts to look the other way, the spot grew larger and larger every day. Then a new crisis arose: one morning my son announced that his tooth was bent. His lower right incisor was now pointing backwards. Not only was it askew, it was also loose; and the next day it fell out—I had forgotten that children lose their teeth.

All children eventually lose their milk teeth and grow a new set. To me this seems peculiar and somewhat wasteful, but it is something that almost always happens. Losing milk teeth is an ever-present step in the sequence of human events leading from baby gums to adult dentition. In logic, we usually think of ever-present members of a sequence as either *necessary* predecessors or as *sufficient* predecessors. Does this logic hold for milk teeth? Are milk teeth necessary and/or sufficient?

We chew with adult teeth 91 percent of our life; if milk teeth are necessary for chewing, then it is only during 6 percent of a lifetime. Moreover, not all adult teeth even have milk teeth predecessors—our molars (which appear at six years in humans, at three years in gorillas, at about one year in other monkeys) have no natural predecessors. Some mammals, like kangaroos, opossums, whales, narwhals, and manatees have no baby teeth (except for two premolars)—instead, they start right in with adult teeth. It appears, then, that there is no a priori reason for mammals to lose a set of teeth and that shedding milk teeth is not a necessity.

If milk teeth are not necessary predecessors, then are they perhaps sufficient predecessors? Does a set of milk teeth inevitably lead to a set of adult teeth, as falling off a log leads to lying on the ground?

Some people never lose all of their milk teeth. (I still have my left front baby incisor, and it works quite well.) These hanger-on milk teeth do not simply overstay their welcome, for if they are eventually lost, there is no replacement by natural adult teeth. Milk teeth are not sufficient predecessors for adult teeth, because when given a set of baby teeth you cannot lean back in your highchair basking in an absolute certainty that someday you will aquire a full set of natural adult teeth.

Milk teeth are neither necessary nor sufficient, and although they are useful they are not essential. Milk teeth are natural and constant members of a sequence of events proceeding resolutely from toothless baby gums to the full set of adult teeth; but there is no logic to milk teeth. Milk teeth are not necessary, and they are not sufficient—they are just *there,* sitting naturally and almost inevitably in mid sequence. In the natural realm, not all pattern predecessors follow the clean, structured, and simple logic of mathematics, because logic is of people, it is not of nature. The biologist must be wary of logical simplicities and clear generalities*—although the need for simplicity draws him back, the complexity of Nature's patterns continually forces the pattern scientist on toward intricate, complex, and convoluted explanations.

---

*Andrew Huxley wrote:

> Let no one imagine that I am denying the truth, or the importance, of generalizations which may be summarized by the phrase 'the Uniformity of Nature'. What I am saying is...that it is usually impossible to tell whether [these generalities] apply to a particular phenomenon until that phenomenon has been thoroughly investigated and explained....[A priori] we may say that the phenomenon is something fundamental and must for that reason be essentially the same wherever it is found, but this argument is merely a definition of the word 'fundamental': if it turns out that the [various occurrences differ, then] it will be said that they were not really 'fundamental' to living matter after all.
>
> [Huxley, pp. 55-56]

PATTERN CONCEPTS

Chapter One

THE LUXURIANCE OF NATURE

Scientists strive for parsimony and simplicity in their abstractions, but Nature is not constrained by these human extremum principles—she is not an engineer. Biologists see this continually, and in an age grown distant from the woods, all pattern scientists should remember that humans are only one of Nature's myriad recurrent and incondensably complex patterns.

When you have eliminated the impossible, whatever remains , however improbable, must be the truth.

[Mr. Sherlock Holmes]

I

Biologists rarely used the word "impossible." To a biologist, the range of the possible is so large and the potential biological entities so overwhelmingly numerous that the impossible is only a tiny issue; it is the wall at the edge of the universe—real and formidable constraints, but constraints lying somewhere far away, somewhere in the realm of the physicist. The biologist's Nature peers at these constraints through the wrong end of a telescope, and the limits of the physical universe form toy fences in another province. With more than 1500 different species of daisies in Europe alone, with more than 2000 different species of crickets world-wide, and with 30,000 different proteins specific to the brain of the rat, the biologist has little room on his desk for perpetual motion machines or for rockets that travel faster than the speed of light.

Biologists rarely use the word "impossible," and in part this is because of the richness of their realm, overflowing with strange variants. In part it is also because the mainstream biological tradition is natural history, the record of Nature's accomplishments, the careful charting of the possible: biologists are artisans, they want to do things and to make things, and they actively seek out the possible—things that can be done and that can be made. Beyond these reasons however, I suspect that there is a more fundamental reason that biologists rarely speak in terms of the impossible. In one sense there are no biological impossibilities: deep within the true province of biology everything is possible.

How can this be the case? Will all things eventually come to a biologist who waits? The biologist is an optimist; he cannot help himself, because the wide-range of the possible in biology is almost the definition of the province of the living. Living beings form a special realm of science, a realm filled with collages of butterflies, grasses, mildewing molds, people, bears, whales and bats, eggs and embryos, grandmothers and grandfathers. Biology is about life, and life is luxurious complexity.

Life is complex organisms. We know them well, these organisms; we pass them every day as we walk on the grass, under the trees, past the birds and the squirrels. Gardening, we run our hands through the earth with its crumbles of plant detritus, worms, larvae, hundreds of microorganisms. We know organisms by touch, taste, and smell; and those organisms that never come within our grasp, such as gulls at the shore, we know by sight and by sound. Best of all we know human organisms; we know our parents and our friends; we know strangers on the bus, on the street, and in the stores; and we know ourselves. How many times a day do we look at our hands? A hundred? A thousand? We feel our toes from outside and from inside, we hear our heartbeats, we smell our sweat and taste our blood. We are not all physicists or economists or mathematicians, but we are all biologists.

Biology is the province of organisms—yogurt cultures and yaks, hawks and humans. These organisms fall into a few general types but into an innumerable number of specific varieties—in fact, each organism is unique. Human identical twins differ in many ways; for example, their fingerprints are different. Even a pair of cloned organisms is sufficiently complex so that during their creation the stochasticism of the world can insinuate itself and change a molecule here or an organelle there and produce two individuals that are ultimately distinct.

If there *is* a constant in biology, it is the rich variety that drives the theoretical biologist to remark continually: "Well, I never would have predicted that!" A major contributor to this rich unpredictability of living things is unnecessary complexity: biological systems often contain more machinery than is minimally necessary to make them work properly. For instance, the major role of DNA is to code for the production of proteins, yet 99 percent of the DNA in human cells does not code for any proteins. Cells in related species of animals can differ in their DNA contents by greater than 100-fold, and this suggests that if the excess noncoding DNA plays any role at all, then it must be exceedingly modest. There is no simple rule for predicting how much DNA a cell will have—the actual DNA content of any particular cell is probably an accident of history, and the theoretician who attempts to construct an a priori table of the DNA contents of cells sets forth a futile exercise.

It is likewise for blood clotting—when you cut your finger, blood proteins clump together, the wound is soon dammed up, and the cut stops bleeding within five to ten minutes. To staunch the blood flow, the initial injury sets off a waterfall of from 8 to 13 separate chemical reactions in two chains, each chemical transformation giving rise to the next in an orderly sequence. Thirteen different proteins (coagulation factors) form the normal clotting cascades in humans, and if one of these factors is missing, the person can have a bleeding disorder such as hemophilia. The complete blood clotting cascade is complex, and a theoretical biologist would be hard pressed to predict its actual details from a priori considerations, from first principles, or from the requirements of blood clotting systems. One of the factors—Hageman Factor or Factor XII—even appears unnecessary: those people who (through genetic disorder) develop without any Factor XII do not have bleeding problems; and, whales, dolphins, and porpoises normally have no Factor XII, yet they regularly survive injuries.

Such rich and excess complexities permeate life. At the molecular level, there are the "futile metabolic cycles" in cells (for example, the opposing phosphofructokinase- and hexosediphosphatase-catalyzed reactions), circular chemical reactions that go back and forth producing and unproducing the same molecules while depleting energy stores to no apparent purpose. There is also the fact that cells can use a variety of alternative fuels. The brain normally runs strictly on carbohydrates, burning the elemental sugar glucose for energy. During starvation, however, the brain takes advantage of these parallel metabolic pathways, switching over to burning ketones which are derived from fats. The chemical pathways of cells are not lean and geometric—they are varied, interdependent, multidimensional, and often overdetermined.

Complex and elusive intricacies also characterize the tissue level of biological organization. The corpus callosum is one of the largest bundles of axons in the human brain, and although it extensively interconnects most areas of the cerebral hemispheres, its functioning is sufficiently subtle that for years no one understood exactly what it does. The corpus callosum is found only in the placental mammals: other mammals (such as opossums and kangaroos) and all other animals live quite happily without a corpus callosum. Moreover, the five out of a thousand individuals born without a corpus callosum cannot be distinguished easily from those people with a corpus callosum It was only through a specialized series of psychological experiments that Roger Sperry finally showed how the two halves of the brain normally use the corpus callosum as their most intimate route of self-communication.

From his desk, the theoretical biologist could not determine the role of the corpus callosum with certainty, and he could not predict its appearance or its use in those animals (the placental mammals) that have acquired it during the last 200 million years. Who could have imagined that the human brain contains two separate minds, a right mind and a left mind, each localized in one of the major cerebral hemispheres? Normally, the two minds are in such close touch that they think alike, they trade thoughts instantaneously, they share the same sensations and emotions, and they act as one. All this intimacy flows through the corpus callosum, yet at the same time each separate brain is a powerful and complete mind. Normally, two brains make each human, and two brains are a wonderful but unnecessary complexity.

The corpus collosum is not a necessity, but is it just a frill? The nematode worm *Caenorhabditis elegans* has two genes for the single enzyme acetylcholinesterase; both genes are expressed in almost every cell in the body, and if one gene is lost the animal is functionally normal. Is one of the genes a frill? We have two brains, two kidneys, two tonsils, two testes, two nostrils—is the second one a frill? Or consider the excess cells that are generated in embryos. In the normal embryonic spinal cord, there is regularly a 100 percent overproduction of young neurons: only half the original complement of cells survive to adulthood, the excess cells die and disappear. Is this purposeful, or is this wasteful? Such questions—questions with words like "frill," "purpose," or "necessity"—are slightly askew. They are difficult to answer because they are built from peculiarly human judgments. Nature need not adhere to human standards, and she need not follow human principles. Nature does as she does, and we can be secure in our science only when we act as natural historians, conscientiously describing the natural realm retrospectively. We can rarely be confident armchair theoreticians, and we walk a precipitous course when we attempt a priori evaluations based on our own anthropocentric standards.

II

The complexity and the individuality of its particular items of study—organisms—are characteristics of the biological realm, but these features do not completely distinguish biology from astrophysics, metallurgy, or economics. Galaxies are complex and individual, alloy metals are complex and individual, the European Common Market is complex and individual; astrophysics, metallurgy, and economics continually face objects of study that have many interactive parts forming unique and distinct wholes. Yet we are rarely confused, we know quite confidently which items of study fit with which discipline. True, the edges of biology overlap other fields—there is a biophysics, a biochemistry, and probably a bioeconomics, but there is also a clear-cut central biology, and the biologist settles comfortably into the center of his own special niche as if it were a well-worn armchair. He may become a little nervous when measuring the refractive index of a lipid bilayer, and he may feel that he is treading somewhat afield when he calculates the cost:benefit ratio for parental care in a chimpanzee troop. Nevertheless, when he catalogues the larval stages of a grasshopper or when he dissects the neuronal architectures of the human brain, the biologist knows that he is truly immersed in his own special element. Something there is that uniquely determines this discipline of biology—something special about the manifold complex patterns of organisms.

What is this special quality? Organisms are patterns of matter that are at once complex and individual, and the features that distinguish these patterns from the other complex and individual patterns of the natural world are the two sequential processes that produce organisms: ontogeny and phylogeny.

Ontogeny is the developmental sequence of an organism, a playscript with all of the stages in the transformation from an unspecialized embryonic form to a particular and idiomatic machine. Ontogenies come in all shapes and sizes. At one extreme, bacteria go through an ontogeny that is entirely internal: the transformations from a single parent cell to two daughter cells are all cascades of changes of molecules inside the cell. On the other hand, multicellular organisms (such as squid, butterflies, and people) begin as single fertile cells and then transform into unified collections of millions of cells. The ontogeny of a multicellular organism is a cascade of intracellular, cellular, and extracellular changes establishing whole cities of specialized ceils. During these ontogenies, interactive pockets of cells are geographically segregated into organs and tissues connected by highways of nerves and vessels; the construction of these cities is continuously dynamic, and it proceeds inexorably in a particular sequence, the characteristic ontogeny of that organism.

An ontogeny is determinate, stereotyped, and highly reproducible. It is like a phonograph record—when conception sets the needle in the first groove, it plays out the full music of a life. Random dust will change the notes a bit, the environment may distort the sound, and occasionally the needle gets caught by a scratch and falls into the endless loops of a cancer. Ontogenies are dogged things, however, and a scratch or even a jarring of the turntable often cause only a skip in the sequence as the needle falls into a different groove and proceeds once again resolutely on its inevitable path.

Besides ontogeny, there is also phylogeny—the ancestral lineages of organisms. Ontogenies are the life histories of individual organisms, and phylogenies are the repeated unfoldings of ontogenies: generation after generation organisms beget like organisms, and our ancestors are our phylogeny. Biological time is different from physical time; the organismal clock ticks in generations. The physical time scale of phylogenies ranges through six orders of magnitude for the same unit of biological time. In 100 years, a human phylogeny contains five generations, a buttercup phylogeny 100 generations, a fruit fly phylogeny 2500 generations, and a bacterial phylogeny as many as three million generations. Thus, a human phylogeny of five generations is a very short organismal time span, and in the course of one hundred years humanness is preserved—although we look somewhat different from our parents, there is always a strong family resemblance. The striking similarities between ancestors and descendants transcend handfuls of generations; those differences that do show up during a few generations are rather subtle, and each child is much more like his parents than he is different from them. To Nature, "time" means "change," and a few generations offer little opportunity for fundamental organismal change. A hundred thousand generations, however, is another story.

Two hundred thousand generations ago humans were not human. At that time, our ancestors were four feet high ape-like creatures living in what is now eastern Africa; they were primates not taller than our children and certainly not as smart, they lived in forests, and they ate roots, tubers, and grubs. Although these animals—*Australopithecus afarensis*—looked much like apes, they walked upright and they differed structurally from their neighboring apes. The notable organismal changes happen on a time scale of hundreds of thousands of generations, and 100,000 generations after *Australopithecus afarensis*, the new members of the human phylogeny *(Homo habilis)* had acquired the large brains that are characteristic of today's humans. In the course of thousands of generations, the differences between members of a phylogeny can become so marked that the original organism evolves into a new organism, and during 100,000 generations *Australopithecus* evolved into *Homo*. If we step back far enough, if we take a patient mountain range view or, better yet, an astronomical view of biological time, we can see that phylogenies evolve, and evolving phylogenies also characterize the biological realm.

One distinguishing characteristic of the biological realm, then, is ontogeny, the recurrent stereotyped re-creation of a very complex pattern, the determinate production of complexity. The other distinguishing characteristic of the biological realm is phylogeny, the evolving ancestral lineages of these ontogenies. The uniquely biological phenomena of the world are complex patterns with ontogenies and phylogenies. Life is not a DNA molecule or a nerve cell or a kumquat or a Wolf spider; life is not a particular thing at a given time. Life is a sequence of autonomous and recurrent stereotyped recreations of certain very complex patterns—life is a child growing and becoming a mother and eventually a grandmother.

III

With Nature's unconstrained waywardness, ontogenies and phylogenies have become highly textured and quite varied. Nature is luxuriant, and her pattern-assembly sequences are not limited to the simplest or the most efficient paths. In the natural realm, organisms are not built by engineers who, with an overall plan in mind, use only the most economical and appropriate materials, the most effective design, and the most reliable construction techniques. Instead, organisms are patchworks containing appendixes, uvulas, earlobes, dewclaws, adenoids, warts, eyebrows, underarm hair, wisdom teeth, toenails, freckles, and upright posture. Organisms are a meld of ancestral parts integrated locally, step-by-step, during their development through a set of tried and true ontogenetic mechanisms that ensure matching between the disparate elements, between nerves and muscles and between bones and joints, but that have no overall vision. Natural ontogenies and natural phylogenies are not limited by principles of parsimony, and they have no particular goals. Possible organisms can be overdetermined, unnecessarily complex, or inefficiently designed.

A good example is vitamin C (ascorbic acid)—the popular home remedy for colds—an enzyme cofactor that is necessary for the healthy functioning of multicellular organisms. Most plants and animals can synthesize their own vitamin C, but a few animals (humans, monkeys, guinea pigs, fruit bats, certain fish, and certain insects) cannot make vitamin C and must get it from their food. It is likely that the ancestral organisms of most phylogenies *could* manufacture their own vitamin C and that this ability was then lost during evolution in certain lineages. From a human's engineering viewpoint, it seems surprising that animals would add a dietary constraint by relinquishing direct control over their vitamin C stores. Nature, however, is not an engineer—she is not required to play by human rules or human principles, and she can take any of the available roads. Why have monkeys and guinea pigs lost the ability to make vitamin C, while goats and mice have retained that capacity? In this case, the answer is probably that it was simply an accident of history.

The constraints in building organisms are usually insufficient to limit Nature to only one blueprint, and a wide range of alternative architectures have evolved. There is no one "right" way to build an eye: octopus and human eyes appear quite similar, but the human eye is built exactly inside out when compared to the eye of the octopus. In an octopus, light passing through the lens falls directly on the photoreceptors, while in a human, light must travel through many layers of cells and axons before reaching the photoreceptors, which are themselves pointing the wrong way—that is, toward the back of the eye. Likewise there is no one "right" way to build a wing, and Nature has used a number of radically different designs. bat wings are modified hands, while insect wings are entirely separate appendages.

There is no one "right" way to excrete the effete products of metabolism; for DNA degradation, some animals (like humans and Dalmatians) excrete purine wastes as uric acid, some animals (like mice and turtles) degrade purine further and excrete the wastes as allantoin, while other animals (like fish and frogs) take the degradation process two steps more, excreting the purine wastes as urea. There is no one "right" blood level of salts; for example, the concentrations of blood sodium and blood chloride differ by 100 percent between bony fish (such as the carp) and cartilaginous fish (such as the skates). Even within the same species, the normal range of concentrations of blood molecules can span a 50 to 100 percent difference—in adult humans, the normal range of blood albumin is $5.4\text{-}7.4 \times 10^{-4}$ M, the normal range of blood copper is $1.1\text{-}2.3 \times 10^{-5}$ M, the normal range of fibrinogen is 1.5-3.6 gm/1, the normal range of glucose is $3.6\text{-}6.1 \times 10^{-3}$ M, and the normal range of magnesium is $0.75\text{-}1.2 \times 10^{-3}$ M.

Not only can the concentrations of critical molecules vary, the actual structures of molecules can vary. Although certain parts of a biological molecule are fairly immutable, there is often no one "right" overall molecule. For instance, insulin is an essential protein hormone built of about fifty amino acid subunits. One to three of these subunits differ between the insulin molecules of pigs, cows, and humans; nonetheless, the insulins from pigs and from cows are perfectly acceptable substitutes for human insulin, and both pig insulin and cow insulin are commonly used to treat human diabetes. For insulin as for other biological entities, many viable avenues are open to Nature. Even the elements composing biological molecules can vary; in some organisms, selenium can be partially substituted for sulphur;

> plants can synthesise large amounts of seleno-proteins, enough to poison stock and to serve as 'systemic' poisons for insect pests, without harming the plant in any way. Similarly the bacterium, *Escherichia coli*, can synthesise seleno-proteins which appear to function as well as the normal sulphur-proteins...
> [Also,] niobium [can substitute] for vanadium in the blood pigment of ascidians. Here even the individuals of one species may have either V [vanadium] or Nb [niobium]...
> In bone, strontium can take the place of calcium, and in fact many ions of either charge can be incorporated into the bone mineral without ill effect...
>
> [Needham, pp. 494-495]

These particular details of an organism are the products of evolution, and evolution has followed the exigencies of the times under the whims of chance and accidents of history. Had the continents not drifted apart, Australian fauna and flora would undoubtedly be less peculiar; Australia would probably have had indigenous hoofed animals and indigenous apes, animal groups that never developed on that island. Had Alexander Fleming not discovered penicillin, penicillin-resistant bacteria would be a freakish oddity rather than ubiquitous coinhabitants of our planet. Or bones—they

> might have been silica...or aluminum, or iron—the cells would have made it possible. But no, it is calcium, carbonate of lime. Why? Only because of its history. Elements more numerous than calcium in the earth's crust could have been used to build the skeleton. Our history is the reason—we came from the water. It was there that cells took

> the lime habit, and they kept it after we came ashore.
>
> [Eiseley, p. 6]

Extant organisms are legacies of the habits acquired by their ancestors, but these habits coalesced from a plethora of possibilities.

IV

Science fiction comes in two varieties. On the one hand, there are the tales that explore worlds harboring phenomena that scientists think are impossible. These stories ask the questions of dreams: How would people spend their evenings if everyone had a perpetual motion machine in his basement? How soon would you get bored if you lived forever? What is on the other side of the wall at the far edge of the universe? On the other hand, there are the science fiction tales that explore worlds that just might exist. These stories ask the questions of science: In what language could we talk to an extraterrestrial creature? What could we do with self-reproducing automata? What will people do when the sun goes out? The standard science fiction of biology falls into both of these categories, but which biological tall tales are the stories of dreams and which are the stories of science?

Consider mushrooms—"the elf of plants" Emily Dickinson called them. Actually, mushrooms are only distantly related to plants; mushrooms are many-celled fungi, relatively advanced organisms with cell walls (like plants) but with no ability to manufacture their own food (for example, no photosynthetic machinery), with no ability to move, and with no nervous systems. We are all aware that many animals are fungivorous, but it took science fiction to popularize the idea that mushrooms could be carnivorous. There are, for instance, the mushroom people of the mysterious planet Basidium-X who must eat chicken eggs to remain healthy [Cameron]. The stories of dreams? No longer, for carnivorous fungi actually exist here on Earth: certain woodland toadstools trap and eat worms, and from worms it may be only a small step to chickens. Carnivorous fungi once in the realm of the tall tales of science fiction are now unequivocally science fact.

Or there is the square organism, once a creature known only in Flatland, where "all Professional Men and Gentlemen are Squares" [Abbott]. Today, the square organism has found a home on our well-worn Earth, swimming in brine pools of the Middle East—there, tiny flat transparent bacteria "in the form of a thin square sheet" [Walsby] float like ghostly salt crystals, mimicking the perfect planar polygons and belying the notion that, to reduce their surface-to-volume ratios, cells must be spheres.

Carnivorous mushrooms and square bacteria bring a smile to the biologist; they are surprises, but only small surprises. They elicit a "What will Nature think of next?" but they do not shake the roots of the biological sciences. Carnivorous mushroom and square bacteria do not stretch the bounds of biology, because they can be explained by mechanisms that sit somewhere in Nature's cluttered shelf of standard organismic machinery. The biologist may have to root around a bit among the everyday mitochondria, the familiar Krebs cycles, and the mundane cyclic AMPs to retrieve all the appropriate mechanisms—and he will certainly have to spend some time in serious study to find how these mechanisms have been stuck together in the peculiar combinations that make carnivorous fungi and square bacteria. Nonetheless, somewhere in a corner of her cupboard, Nature is sure to have just the right bits and pieces to construct these natural oddities.

Nature regularly builds baby carnivorous fungi from spores of parent carnivorous fungi and replicates daughter square bacteria from parent square bacteria, and Nature derived the parent carnivorous fungi from other preexisting fungi and the parent square bacteria from other pre-existing bacteria. Strange as they are, carnivorous mushrooms and square bacteria—the incarnations of biological tall tales—are neither ontogenetically nor phylogenetically impossible. But how about angels? Although they adorn the spiritual world, our biological world has no angels. Why is this? The answer is that, while it is not absolutely impossible, it is nonetheless quite difficult for Nature to construct an angel in an extant phylogeny. In addition to the arms and the legs of a human, an angel has a set of wings along its back. Wings are complex structures sculpted of muscles, bones, and nerves, and an angel's wings are covered with feathers. To introduce wings or any other complex appendage into an extant phylogeny, Nature needs the appropriate raw materials and the appropriate organizational blueprint, but in this case the pre-existing structures are not yet available for angels.

The wings of today's flying vertebrates (the birds and the bats) are direct modifications of preexisting front limbs; the muscles, bones, and nerves were already there in ancestral organisms, and Nature evolved wings by stretching, shrinking, folding, and bending these preexisting elements. Through all of the geometric distortions, the overall organization—the topology—of the front limb has remained the same. (For example, the upper limb always has a single long bone, the humerus, and the lower limb always has a pair of parallel long bones, the radius and the ulna.) In contrast, the back of contemporary mammals has no preexisting structures that can be stretched or shrunk, folded or bent, into a wing. To make an angel today, the fundamental ground plan—the topological arrangements—of the existing elements must be tampered with, and new structures must be generated without precedent. This Nature cannot easily do, because she "behaves like a tinkerer," [Jacob] not like a creative engineer. Thus, for the near future, angels will probably remain spiritual and winged horses like Pegasus will remain mythological. But biologists do not consider these creatures impossible. Instead, angels and winged horses are put into another realm: they are highly improbable biological phenomena. Improbable biological phenomena cannot easily be pieced together by Nature from any of the mechanisms in her crowded cupboard of organismic machinery and in the context of extant phylogenies. Without the coincidence of a number of highly improbable events—arcane and wondrous initiating events—Nature cannot generate an angel or a winged horse via an existing ontogeny.

But here is where the biologist unveils his soul. The biologist is a scientist who luxuriates in complexity, and in his home, buttermilk-thick with life and under the patience of millions of generations, Nature sometimes stumbles on the extremely improbable: sometimes arcane initiating events can indeed be quite natural. Arcane events just "happen to happen" and are not very likely to happen again, and when they arise in a natural ontogeny or phylogeny, arcane events can trigger sequences that are as natural as apples. It is the initiating event in the generation of an angel that would undoubtedly be arcane, but the autonomous development of an angel from a tiny wisp of an angelic embryo or the spontaneous unfolding of a lineage of angels, once set on their way, would become natural phenomena no more impossible in the biological realm than the autonomous development of an oak from an acorn or the spontaneous unfolding of the ancestral lineage of the great Bach family.

Physically impossible organisms, such as hedgehogs that run faster than the speed of light and perpetual motion bees, can be dreamt by the physicist, but I cannot easily imagine a biologically impossible organism. When we have eliminated the physically impossible, when we remain within the constraints set by the physical limits of the universe, then whatever remains—no matter how improbable—must be considered biologically possible. For Nature is wild, luxuriant, and rich, and her splendor is unconstrained—and afternoons poking about the Woods Hole seashore among the horseshoe crabs, the seaweed, and the tunicates, or munching blue-eyed scallops and beach peas on a rocky island in Penobscot Bay, or chipping ornate brachiopods from the shale of the Chagrin River make me hesitate to think that I could ever dream of a creature that might not creep out from among the cattails one cool and misty springtime morning.

Chapter Two

SIMPLE AND COMPLEX

Chapter Two

SIMPLE AND COMPLEX

The simple is bare and pure; it is single, uncompounded, and of uniform structure. In contrast, the complex is ornate and mixed; it is intricate, involved and of diverse structure. This has been the classic dichotomy, but the computer age has gone beneath this fundamental split—computer theory has given us a new insight into the essential difference between simple and complex. Machine-based definitions emphasize that the complexity of something is directly proportional to the difficulty of its fabrication; in turn, the difficulty of fabrication can be measured by the magnitude of the necessary precursor components. Thus, simple patterns can be built from small beginnings, but complex patterns need extensive and manifold origins.

We can see some of the implications of these definitions when peering with steady gaze through our adult eyeglasses, we can hear some of the implications in the even cadences of our scholastic colleagues, but the world is a large place and it has other eyes and other ears—and especially there are children. Children see with a fresh-born perspective, they hear with a crisp musicality, and

> *then too they sing, only you must listen with your littlest and newest ears if you wish to hear their singing...Each song is softer than an eye wink, softer than a Nebraska baby's thumb.*
>
> *[Sandburg, p. 206]*

For a full, special, and clean vision, we should talk to our children; we should undertake the challenge of building our arguments, our reasoning, and our understandings in the most elemental and pristine terms. And, what do you say to your six year old when she asks you the difference between "simple" and "complex"? Ah, what do you say, and what do you say, and what do you say indeed...

I

The world is filled with many things.
Some are simple—straight white strings.
Some are more complex instead,
Like twisted, tangled yarn and thread.

A ball is simple, round and calm.
You can hold it in your palm.
And in one hand you carry it free
And unencumbered, easily.

And if it's small, then you can slip
It in your shirt and lightly flip
It out when you're inclined
(Like when a ball game comes to mind).

But a pile of puppies is complex—
One jumps up and then the next.
They wrestle in a wriggling bunch;
You cannot see them all at once.

You cannot hold them in your hands
(Not all at once)—like fluffy sand
They slip and slide and squirm around
Trying to lick you or get down.

They'd never sit all quiet inside
A package waiting for a ride;
For bunched together in a sack
They'd scratch and tickle, scramble back

And tumble out onto the floor.
A ball is simple, but puppies more
Than make a pile—they are complex,
All jumbled feet and ears and necks.

A ball in pocket's out of mind
It's stored, forgotten—quiet, resigned;
But you can't forget a complex store,
A sack of pups you can't ignore.

II

What other simple things are there?
To what complex do they compare?
A school of minnows neat and dense
Is like a simple picket fence.

But writhing in a scramble trembling
Octopuses, all resembling
Tangled barbed-wire broken fence,
Are complex, twined with little sense.

Simple is quite plainly built,
But complex is a crazy quilt.
Simple is shaped by easy rules,
Complex things need many tools.

Little families are a simple mix:
A hen, a rooster, baby chicks;
Or mother, father, baby child—
No complex relations, mazelike, wild.

But--mother, father, baby, *and*
Brothers, uncles—the whole clan
Of sisters, cousins, nephews, aunts,
Grandmas, grandpas, potted plants,

Nursemaids, stepsons, marmosets,
Nieces, godsons, still more pets,
Like dogs and goldfish, hamsters, fleas—
These form complex families.

III

Simple things have simple parts—
Few different arms and legs and hearts.
Their parts are even, much the same,
They look akin and fairly plain.

Take water—it's a simple stuff—
All drops of water alike enough.
(You'll politely address a drop named Fred
Then find you've talked to Ted instead.)

Gumdrops, bubbles, spoons and wheels,
Salt and sugar, prisoners' meals,
White thumbtacks and plastic bowls,
Sand and vacuum, doughnut holes—

Simple things are uniform,
But complex things are like a storm.
Their parts are different, mixed together—
Underbrush in rainy weather,

Jungle vines and seaweed mats
And intertangled sewer rats,
Butterfly wings and Easter hats
And Christmas trees and calico cats,

Computers, jet planes, city streets—
Complex things are what one meets
In forests, oceans and the zoo
Where each thing can be somewhat new.

Many parts make complex things,
With many arms and many wings—
Peninsulas and bays and capes
In strange surprising wondrous shapes.

Complex things can each one be
Special and different, like each tree
Inside a woods. Each is complex,
Each is new, each slightly different from the next.

IV

Stories can be simple tales
Or else complex, with bears and whales
And dragons, spaceships, nighttime raids
And crowds of people on parade.

Or poems—simple ones are smooth and free;
Their pictures flash up instantly:

"Rain on the green grass
And rain on the tree
Rain on the housetop
But not on me!"

Simple is easy to command,
You read it once and understand.

But complex poems seen in a trice?
No, one reading will not quite suffice:

"Rain down, pattering
On leaf stick creek
Wet round up drops
To wink by cheek"

Full of parts to wonder about,
A complex poem takes thinking out.

V

People are complex—think of you,
With varied parts and colors too:
Wrists and toes and ears and cheeks,
Elbows, knuckles, hair and teeth,

Rounded knees and squarish chin,
Oval eyes and fingers thin,
Bumps and spots, bulges, bends—
You're different from your many friends.

You're distinct, complex and rich
And multiform, like snowflakes which
Are each unique—through people, rife
Diversity is brought to life.

Your smile, your laugh, your fingerprints,
Your eyes, your voice, your hair's full tints—
You're so complex, no child so fine
Is just like you, wee girl of mine.

Chapter Three

'TIS THE GIFT TO BE SIMPLE

Scientist wrestle with natural patterns, trying to understand the simple and the complex orderly assemblages with which Nature has populated our universe. In the end, we humans comprehend simple things best, and Physics has claimed that corner of the universe where natural simplicity reins. Biology has a far different realm; there, truly complex phenomena—Nature's incondensable complexities—temper the pattern scientist's inborn hope for simple explanations.

I

In the sciences, physicists are the chosen people. The great laws of physics are beautiful and elegant, and physicists see the world through crystalline lenses of mathematical simplicity.

I envy the physicist. Following in the Pythagorean and the Platonic traditions, he has built his natural laws from a geometrically simple model of the world. For him, the universe is elegant, mathematically accessible, and numerically tractable, and this is reflected in the many small and direct formulas that provide the foundations of physics. Often originating as explanations of individual examples, these simple formulas (such as $F=ma$ and $E=mc^2$) have become handy pocket-size summaries of innumerable specific cases: each short equation stands proxy for an unlimited number of particular real-world applications. Ad hoc mathematics based on the principles of simplicity worked repeatedly for the physicist, and the continual successes of elegant mathematics seemed too consistent to be explained by mere chance—how can the rules be so simple unless the universe itself is inherently simple? As the physicist Eugene Wigner wrote: "It is difficult to avoid the impression that a miracle confronts us here." [Wigner, p. 229]

The miracle of physics, however, is only an atheist's wonderment; it does not puzzle the devout scientist. In a more religious age, the orderliness of physics was not a surprise. Three hundred years ago, the mathematician and logician Gottfried Wilhelm Leibniz pointed out:

> It follows from the supreme perfection of God, that in creating the universe he has chosen the best possible plan,...the best arranged ground, place, [and] time

For example,

> The supreme wisdom of God led him to choose the laws of motion best adjusted and most suited to abstract or metaphysical reasons...And without this it would not be possible to give a reason why things have turned out so rather than otherwise.
>
> [Leibniz, pp. 528-529]

At one time, scientists openly presumed that, God being perfect, the laws of nature must therefore be parsimonious and simple. Leonardo da Vinci said: "every action done by Nature is done in the shortest way," and Isaac Newton's *The Rules of Reasoning in Philosophy* began:

> We are to admit no more causes of natural things than such as are both true and sufficient to explain their appearances . To this purpose the philosophers say that Nature does nothing in vain, and more is in vain when less will serve; for Nature is pleased with simplicity, and affects not the pomp of superfluous causes.
>
> [Newton, p. 202]

And, although it is not written explicitly into contempory science, the physicist's God—the God "Who," as James Weldon Johnson [p. 20] wrote, "lit the sun and fixed it in the sky,/Who flung the stars to the most far corner of the night,/Who rounded the earth in the middle of His hand"—is not an agnostic's God. Instead, He remains a God Whose laws man might comprehend and mirror in clean, clear equations: "Our experience up to date," wrote Albert Einstein, "justifies us in feeling sure that in Nature is actualized the ideal of mathematical simplicity." [Einstein, 1934, p. 167]

The physicist's God has forged a neat, tidy, and parsimonious universe. If at times the scientist's skein of data appears tangled, the order messy, the organization imperfect, and the synthetic principles fuzzy and unraveled along the edges, then it is only the burden of our imperfect and somewhat scattered human insights. We humans have an intemperate and hasty need to know everything at once, and to the physicist the apparent complexities in the universe stem from our incomplete or disorganized knowledge. Human intemperance complicates nature, says the physicist. It is the intemperance by which Eve ate the original apple of knowledge and was cast from the simple joys of Eden into the complexities of the real world. With patience, persistence and insight, the physicist hopes to regain the well-tempered life, to rediscover the simpler, brighter, clearer truths of our childhood in Eden: in his heart, the physicist believes that the world is truly and elegantly simple.

Like physicists, biologists have looked for a tidy and well-ordered universe; we too want a universe organized by simple principles of harmony and perfection. Before Darwin, biologists contended either that animals are already perfectly harmonious creations (Galen said that in organisms "there is nothing ineffective or superfluous or capable of being better disposed") or that animals had been set upon a preordained course toward a flawless elegance; for instance, the Swiss naturalist Charles Bonnet wrote (1769):

> animals are called to a perfection, for which the organic principle existed at the creation, and whose complete development is reserved for the future state of our globe.
>
> [Bonnet, cited in Gould, p. 25]

After Darwin, the perfecting mechanism changed but the principle remained the same:

> As natural selection works solely by and for the good of each being, all corporeal and mental endowment will tend to progress towards perfection.
>
> [Darwin, p. 373]

These were hopeful declarations—they posited a clear and rational natural realm. Unfortunately, biologists' hopes have continally been confounded by the fact that organisms are in reality a messy, complex, and Byzantine lot. The ancient miracle of elegance and simplicity in physics did not completely overflow into the other sciences, and biology has been the notable exception.

Biology has never fit comfortably into a world of simple crystalline order. Organisms are heterogeneous and complex at all levels, whether we look at them from a distance or study them through a microscope. At the species level nature bubbles with varied creatures: snakes, sloths, and snails; flatworms, flatfish, and flat peas; sponges, paramecia, and bacteria. And what are we to make of the tribes of aspiring organisms, the stunted creatures—mitochondria and chloroplasts, which are now parasitic organelles—and the viruses, episomes, plasmids, and prions, the mobile replicating flotsam and jetsam of cells? Biologists have tried to force a clean, simple, and even mathematical order among such organisms (Figure 3.1), but family trees remain complex.

In part, this complexity stems from numbers: today, taxonomists recognize more than one million species of animals (three-quarters of which are insects) and four hundred thousand species of plants. In addition, the complexity comes from organismal histories: Nature has not been limited to any one avenue of evolution, and the taxonomist must be an empirical historian. Which of the many possible family trees is the particular one that actually unfolded during past eras? What are the real historical relationships? Currently, taxonomists debate whether the insects and the scorpions are descended from common arthropod ancestors; they worry whether amphibians arose twice independently—did newts and frogs evolve from two distinct populations of fish rather than from a common stock of amphibians? They are unsure of elemental anatomic homologies—is the endostyle of the sea squirt a primitive version of the thyroid gland of the human? They are puzzled by apparent *ex nihilo* evolutionary innovations—where

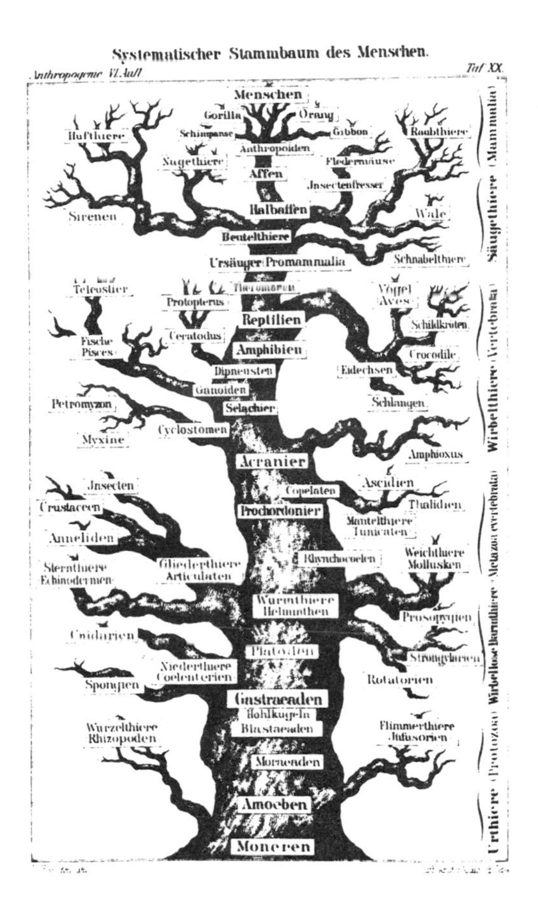

Figure 3.1

At the end of the last century, the German biologist Ernst Haeckel aimed for an elegant "crystallography of organic forms." His phylogenetic tree of life neatly ordered all creatures, ascending from the lowly ***Monera*** (prokaryotes—cells without nuclei) at the base, to ***Menschen*** (humans) at the crown.

did the corpus callosum suddenly come from in those mammals with placentas? These questions cannot be answered on the basis of first principles or by a priori reasoning. Nature's answers are idiosyncratic, and we learn them with confidence only through retrospective natural history.

With a haze upon their family trees and a roiling diversity in their genealogical branches, the variety among creatures perplexes even devout simplificationists. Organisms range in size over nine orders of magnitude; lifespans fill five orders of magnitude. Physiological parameters vary widely with no simple rules—the average body temperature in mammals ranges from 23 °C in the spiny anteater to 40 °C in the domestic goat, and body temperatures can drop almost an order of magnitude to 4 °C in hibernating mammals. (Moreover, environmental temperature need be no obstacle; bacterial spores will survive for hours in boiling water, and the failure of heat to kill spores of the bacterium *Bacillus subtilis* in hay prolonged into the last half of the nineteenth century the defense of the Spontaneous Generation of Life.) Diversity is also the rule within organisms: humans have greater than 200 clearly distinct cell types, and each cell type comes in many forms. Neurons, for instance—a single human brain has a trillion neurons, which take on hundreds or thousands of distinct sizes and shapes within the same individual; and neurons can be three thousand millimeters long in the leg or only one half millimeter long in the eye.

II

Our day-to-day experiences make the diversities of animals and of cells feel natural: people are all different, no two houseplants are identical, my golden retriever has (to my ear) an absolutely distinctive bark. Individuality and variation are the birthrights of organisms, and real as is the complexity at these levels, we have nonetheless grown accustomed to it. Molecules, however, are closer to the hard-edged chemistry of precise science, and the extreme molecular heterogeneity of life can be more jarring.

Proteins are the most common organic molecules, and Nature manufactures billions of distinct proteins, completely filling a size spectrum from the small polypeptide chains such as insulin (about fifty amino acids long) to the large polypeptides such as the muscle protein myosin (about eighteen hundred amino acids long). A dramatic microcosm of life's molecular complexity is played out through the antibody proteins of the mammalian immune system. Our large repertoire of antibodies is generated through the shuffling of a pool of interchangeable genetic units, and from a modest genetic endowment the B-cell lymphocytes of one individual can reproducibly manufacture millions of different antibodies. When taken as a whole, the antibody repertoire of an individual forms a molecular fingerprint—no two people have exactly the same protein composition.

Proteins give life its extensive heterogeneity, and all proteins are coded for by the DNA of the genome. In this way, the detailed information for building unique organisms can be traced back to the DNA molecules that are the genes. DNA is the ubiquitous stuff of inheritance, and the genetic code is universal; nonetheless, the scope and the variety of natural DNA sequences are overwhelming. Most devastating to the simplificationist, however, has been the discovery that many of these genetic sequences are so complex as to be indistinguishable from random (Figure 3.2). As the molecular biologist Jacques Monod wrote:

> The first description of a globular protein's complete sequence was given by Sanger in 1952. It was both a revelation and a disappointment. This sequence, which one knew to define the structure, hence the elective properties of a functional protein (insulin), proved to be without any regularity, any special feature, any restrictive characteristic. Even so the hope remained that, with the gradual accumulation of other such findings, a few general laws of assembly as well as certain functional correlations would finally come to light. Today our information extends to hundred of sequences corresponding to various proteins extracted from all sorts of organisms. From the work on these sequences, and after systematically comparing them with the help of modern means of analysis and computing we are now in a position to deduce the general law: it is that of chance. To be more specific: these structures are "random" in the precise sense that, were we to know the exact order of 199 residues in a protein containing 200, it

A-T-G-G-C-T-C-T-C-T-G-G-A-T-C-C-G-A-T-C-A-C-T-
G-C-C-T-C-T-T-C-T-G-G-C-T-C-T-C-C-T-T-G-T-C-T-
T-T-T-C-T-G-G-C-C-C-T-G-G-A-A-C-C-A-G-C-T-A-T-
G-C-A-G-C-T-G-C-C-A-A-C-C-A-G-C-A-C-C-T-C-T-G-
T-G-G-C-T-C-C-C-A-C-T-T-G-G-T-G-G-A-G-G-C-T-C-
T-C-T-A-C-C-T-G-G-T-G-T-G-T-G-G-A-G-A-G-C-G-T-
G-G-C-T-T-C-T-T-C-T-A-C-T-C-C-C-C-C-A-A-A-G-C-
G-C-G-A-C-G-G-G-A-T-G-T-C-G-A-G-C-A-G-C-C-C-C-
T-A.....G-T-G-A-G-C-A-G-T-C-C-C-T-T-G-C-G-T-G-G-
C-G-A-G-G-C-A-G-G-A-G-T-G-C-T-G-C-C-T-T-T-C-C-
A-G-C-A-G-G-A-G-G-A-A-T-A-C-G-A-G-A-A-A-G-T-
C-A-A-G-C-G-A-G-G-G-A-T-T-G-T-T-G-A-G-C-A-A-T-
G-C-T-G-C-C-A-T-A-A-C-A-C-G-T-G-T-T-C-C-C-T-C-
T-A-C-C-A-A-C-T-G-G-A-G-A-A-C-T-A-C-T-G-C-A-
A-C

Figure 3.2

The genetic code for insulin.

The amino acid sequences of many molecules show no simple rhyme or reason; often they are so complex as to be indistinguishable from random. Insulin, the first protein to be fully sequenced, is a typical example. (The letters symbolize the four different nucleotide bases that compose DNA: A=adenine, C=cystosine, G=guanine, T=thymine—from Perler et al., 1980.)

> would be impossible to formulate any rule, theoretical or empirical, enabling us to predict the nature of the one residue not yet identified by analysis.
>
> [Monod, p. 96]

The complexity of the genome is so great that it is in a fundamental sense incondensable—for this reason, Monod has called the complexity of life "indecipherable."

At face value some biological entities, like brains, DNA, and large human populations, appear to be incondensably complex. But is this really the case? "Incondensable" and "indecipherable" have such finality: they are bleak terms. Are complex natural phenomena forever consigned to such cold fates?

People are undaunted optimists; babies make me happy—I cannot resist smiling—they give me the glimpse of a limitless future, cheating death and promising all dreams. When I see a baby, I wonder whether there still might be hope for the simplificationist. Some time in the future scientists cleverer than we will tackle the difficult subject of theoretical biology—maybe they will discover new and wonderful ways to reduce biology to simple explanations; maybe they will eventually learn to parcel all phenomena into smaller and briefer building blocks, forming concise summaries of even the most complex things. Perhaps simplification will someday succeed everywhere.

In the last century, the confidence of simplificationists was unbounded: all of nature and all of human dreams would eventually be caught and abstracted, and each bit of the universe would then be positioned in a balanced, orderly, and relatively simple constructure of graceful mathematics. By 1900, the great German mathematician David Hilbert had outlined an overall mathematical map of all of logic, and his colleagues and students began to fill in the specific topographic features. Although many details of this exercise remained unclear, the mathematicians believed in their grandchildren.

In 1931, however, this unbridled confidence was suddenly tempered—with Kurt Godel's striking Incompleteness Proof mathematicians were forced to admit that logical constructions have critical and inescapable limitations. All conceivable theoretical edifices cannot be built with elegance, completeness, and precision, and there are fundamental and inevitable constraints on the scope and the depth of any of our abstractions. Time, patience, and grandchildren will not give us limitless power to simplify.

The limits of logic permeate all theoretical science. For problems of complexity, the constraints were first made explicit in the mid-1960's. To explore this problem, the mathematicians Andrey Kolmogorov and Gregory Chaitin began with an ideal computer, a machine that could build any conceivable binary pattern (a string of ones and zeros). Kolmogorov and Chaitin then demonstrated that the vast majority of the numerical patterns that could be formulated by such a machine would be "maximally complex"—that is, incondensable to some simpler form. A maximally complex pattern cannot be summarized or abbreviated, it cannot be more briefly abstracted; it cannot be compressed.

What exactly is maximal or incondensable complexity, and how does it relate to our ability to create simple scientific explanations? We all have an intuitive feel for "simple" and for "complex." A ball, a bat, a baby's smile—these things are simple; a computer, Yosemite National Park, the Manhattan Telephone Directory—these are complex. Our natural understandings of simple and of complex appear to be built into our nervous systems, and each concept already comes with its own special emotional cast. Simple things are bright, fresh, and complete; they have a clean glow. In contrast, I find complex things thick, tangled, and forbidding; they have shadows, and they radiate an unease.

Beyond our emotional reactions, "simple" and "complex" have now also been tied to precise technical meanings. Our innate notions of simple and of complex are based on parts and on interconnections: we judge complexity from internal architecture—by the number and the variety of parts and by the number and the variety of interconnections among those parts. Technically, simple entities have few elements, few differences between elements, few interconnections, and few different kinds of interconnections. In contrast, complex entities are manifold and tangled. In pattern terms, simple things have homogeneous contents and homogeneous configurations, while complex things are heterogeneous throughout.

Consider groups of people. At the simple end of the spectrum is the nuclear family, the extended family is a bit more complex, and cities of people are more complex still. At the farthest end of the spectrum, there is the entire world population, comprising a vast range of people and a confoundingly intricate web of relationships; across the globe, there are almost as many different people and as many different relationships as there are people themselves. For this reason, no simple brief summary statements can embody a complete roster of mankind—to completely describe the world population, one needs a monumental compendium. Summary statements and precis are inadequate, and one could probably do little better than actually to list each individual with all of his specific relationships.

It is precisely in this way that the complete population of the world is incondensably or indecipherably complex: any complete description must include so much individual detail, particular variety and specific examples that it would become a list almost as big as the population itself. The explanation would be as large as the thing to be explained. Although simple things can be briefly explained, complex things require large explanations.

The Manhattan Telephone Directory is complex—in fact, it is maximally complex. No simple formulas or rules can generate the directory; the simplest way to create a Manhattan telephone book is to copy another one. Kolmogorov and Chaitin have shown that a great many patterns are maximally complex and that incondensable complexity is inescapable. Incondensable complexity is inextricably woven into the fabric of the universe. Simplificationists cannot depend on their grandchildren, and some natural phenomena can never be reduced to simple beginnings.

III

Humans live intimately with this pervasive complexity, but we are not at ease with it. Innately, we are simplificationists. Complexity is not the natural resting place of the mind, and this colors our reasoning—for instance, we prefer simple explanations.

In essence, an explanation is translation; it is allegory, parable, and metaphor, rephrasing one thing in other, independent terms. Thunder, for example—during a rainstorm, my two-year old explained to me that thunder is "big wheels rolling around in the sky." Typically, his explanation reconstructed the "explainee" in different and independent units: he translated thunder into other things with which he was already familiar. Explanation recreates something in a different medium.

Explanations are the heart of our transactions with each other. Art is a form of explanation, subtle and refined. In the arts, explanations can be simple, like the nursery rhyme:

> Here am I,
> Little Jumping Joan.
> When nobody's with me,
> I'm all alone.

Or they can be more complex, like e.e. cummings's little poem:

> neither could say
> (it comes so slow
> not since not why)
> both didn't know
>
> exeunt they
> (not false not true
> not you not i)
> it comes so who

In the arts, however, complex explanations are freely chosen, because even when faced with complex phenomena the artist can always propound simple explanations: artists can choose to be abstractionists, and abstractionists have the liberty to simplify complex reality, to rebuild the world in a few simple elements, to actively choose simplicity. Paul Klee has written:

> The legend of the childishness of my drawings must have originated from those linear compositions of mine in which I tried to combine a concrete image, say that of a man, with the pure representation of the linear element. Had I wished to present the man 'as he is', then I should have had to use such a bewildering confusion of line that pure elementary representation would have been out of the question. The result would have been vagueness

beyond recognition. And anyway, I do not wish to represent the man as he is, but only as he might be.

[Klee, p.53]

The man "as he might be" is abstractionism, the selective rendering of only certain aspects of a complex phenomenon. As an abstractionist, Klee was free to explain the world with purity, conciseness, and simplicity.

In contrast, Klee's "bewildering confusion of line" denotes realism, the complete explanation of a very complex phenomenon. Realists attempt to recreate the entirety of something in a different medium, and realists must follow the lead of the world: if the phenomena are complex, so must be the translation.

Science is realism. Artists can choose between simple and complex explanations, but scientists are shepherded along only one path; and for complex things, the scientists' explanations inevitably become rich, thick, and nuance-laden—they are translations as complex as the phenomena to be explained.

For example, there are the complex phenomena of the mind. The psychologist Jean Piaget has built one of this century's most important scientific explanations of the development of the mind. His theory, framed through more than fifty books and hundreds of papers, unfolds cognition in natural and spontaneous steps. Three major periods—the sensori-motor period, the symbolic period, and the operational period—follow embryogenesis, says Piaget, like inevitable geological eras. Each period is subdivided into an orderly sequence of stages; the sensori-motor era, for instance, comprises six stages. Every child goes through these stages in the same sequence; some children hurry through them and some dawdle, but all children must play out a preceding stage before graduating into its successor.

Piaget was a natural historian; he began by directly asking children questions:

Experimenter- "If I give you some coffee in one cup, will it still be the same if you pour it into two glasses?"

Child- "I'll have a little more."

Experimenter- "Where?"

Child- "In the two glasses of course."

[Piaget, 1965, p.8]

Piaget watched children (especially his own three children) and then patiently recorded all the specific acts of mental development that he could see:

> I put a chain back into the box and reduce the opening to a 3 mm [slit]...[Lucienne, age 1 year and 4 months] looks at the slit with great attention; then, several times in succession, she opens and shuts her mouth, at first slightly, then wider and wider! Apparently Lucienne understands the existence of a cavity subjacent to the slit and wishes to enlarge that cavity...Soon after this phase of plastic reflection, Lucienne unhesitatingly puts her finger in the slit and instead of trying as before to reach the chain, she pulls so as to enlarge the opening. She succeeds and grasps the chain.
>
> [Piaget, 1952, p. 338]

Together, the children's answers and Piaget's direct observations form the essence of his explanation—a theory interlaced with a plethora of specific examples. Innumerable case histories give the explanation fullness, texture, and a rich heterogeneity; they present both the details and the variety that are necessary for any full understanding of children's cognitive development and they make the complete explanation long and intricate. The child's mind is a complex thing, and, as Piaget has shown us, any attempt at full explanation soon becomes quite complex in itself.

Like the phenomena that they interpret, such complete scientific explanations fall on a spectrum. At one end lie the simple explanations; these cover simple natural phenomena, such as objects falling in a vacuum. As we move farther from the simple phenomena, simple explanations must be supplemented by additional details and qualifications, specific case histories, and exceptions to the rule. And, at the far end of the spectrum, in the realm of such phenomena as children's minds and evolution and the transformation of a one-tenth millimeter egg into a two thousand millimeter human, explanations are almost entirely exception and no rule. Here lie the unavoidably complex explanations, and in those disciplines such as psychology and biology where incondensable complexity is commonplace, the scientist as realist must rest with complex explanations.

IV

Abstractionists' explanations can be simple, but realists' explanations are as complex as the phenomena they describe—and when they are very complex, explanations become indecipherable or incomprehensible. In the sciences, realism forces complex explanations upon us, and this poses an emotional problem. Maximal complexity is tied to incomprehensibility, and we are left with an instinctual discontent because humans are innately uncomfortable with things we cannot comprehend.

Comprehensibility—"What," asked the musical simplificationist Anton Webern,

> "does the actual word...express? You want to 'get hold' of something [and] if you take an object [entirely] in your hand, then you have grasped it, you 'comprehend' it."
>
> [Webern, p.17]

At the most elemental level, comprehension means being able to hold something entirely in the palm of your hand: it means embracing something fully and simultaneously.

Innately, we understand things in natural whole units, and the most fundamental units are those items we can immediately comprehend. Comprehensible things can be held completely: a solid blue marble, the sudden whiff of a warm pie, the flash of a neon sign—these are snapshots that we catch at once in their entirety. As Francis Bacon, the champion of inductive logic, wrote:

> The human understanding is most excited by that which strikes and enters the mind at once and suddenly, and by which the imagination is immediately filled and inflated.
>
> [Bacon, p. 472]

All of our higher mental activities consist of unconsciously shuttling these instantaneous natural units of comprehension about, weaving them into intricate tapestries. Mentally and automatically, we overlay the snapshots in pairs, and then we rework the juxtaposed images—our large contrasts, judgments, and comparisons are always built from a series of smaller, simultaneously perceived images.

(Digital computers also work by these principles. All basic computer operations are interactions of binary arrays, sets of zeros or ones that are registered simultaneously on some microscopic device; in this way, each array is a snapshot, a simultaneously registered—fully comprehended—collection of binary digits. The arrays may be as tiny as one bit or as large as millions of bits, but in all cases the machines work by sequentially overlaying these simultaneously perceived and comprehended units.)

For humans, our dependence on simultaneously-embraced perceptions is rooted in primal childhood, and it is associated with the strong and warm feelings of our bright new baby world. Piaget has pointed out that all of our earliest perceptions are global snapshots-simultaneously perceived whole images; and imbibing these full perceptions is then the primary act of comprehension. Thus, the holistic snapshot is the elemental unit of cognition, and absorbing such whole images is our first form of understanding. Instantaneous images that are absorbed whole and complete are the foundations of thought, of understanding, and ultimately of science. When all else waivers, when our confidence is shaken in the solidity of other building blocks, when closer inspection shows that our atoms are fuzzy or ephemeral, we can, nonetheless, always rest secure with simultaneously perceived elements. These instantaneous snapshots form the unshakable natural units of logic. Elements that are immediately associated and directly juxtaposed are bonded by simultaneity, the most intimate and stable of ties, and this intimacy gives us a primal sense of comfort. We hold the world in our hand, and we are at ease.

Simultaneous juxtaposition is primordial, and it is the one operation that Einstein considered absolute and indissoluble. As the basis for his Special Theory of Relativity, Einstein proposed that simultaneity is the crucial and inescapable ingredient of all measurements (quantitative comparisons); he wrote, for example:

> All our judgments in which time plays a part are always judgments of simultaneous events. If, for instance, I say "That train arrives here at 7

o'clock," I mean something like this: "The pointing of the small hand of my watch to 7 and the arrival of the train are simultaneous events."

[Einstein (1923), p. 39]

In other words, temporal simultaneity is actually spatial simultaneity *simultaneity,* as the elemental unit of comprehension, is the foundation for the Einsteinian revolution in science.

And it is simultaneity that gives "simple" and "complex" their different emotional casts. Specifically, simple things are comprehensible because they are small enough to be fully perceived simultaneously; Leonardo da Vinci wrote:

> We know well that our sight, by rapid observations, discovers from one vantage point an infinity of forms; nevertheless it only understands one thing at a time.
>
> [cited in Gleizes and Metzinger, p. 14]

Simple things can be held in the palm of your hand, and they can be carried in your pocket—for these reasons, simple things are warm and deeply satisfying. Complex things, on the other hand, require wheelbarrows. Complex things can be too large, too oddly shaped, or too convoluted to be held in their entirety at one time; the full extent of a very complex thing cannot be embraced simultaneously either in our actual eye or in our mind's eye, thus very complex things are not immediately comprehensible. Very complex things are big, thick, tangled, and instinctually dissatisfying; and it is the primitive urge for warmth and satisfaction that drives us toward simple explanations.

We hope for simple explanations, but realists such as scientists cannot escape complex explanations. Realists cannot always recapture the primal satisfaction of simplicity; only the abstractionist can be assured of returning to a simpler childhood. Biologists and psychologists, economists, historians, metallurgists, meteorologists-those realists who trade in the full explanations of inescapably complex phenomena-trudge the wilderness pulling behind them dense and intricately woven theoretical patterns. For realists, complex explanations are sometimes unavoidable.

Physicists have been fortunate: early on, they staked their claim and chose to attend to those natural phenomena that are on the simple end of the spectrum. But physicists are realists, and on the edges of their universe they too face extreme complexity. No scientist is immune-it is only the abstractionist who can perpetually climb back into the Garden of Eden; there, he can gently rock forever in the warm arms of an elegant Lord, the Lord of his childhood, a Lord Whose wisdom knows only an awesome and pure simplicity.

Chapter Four

RANDOMNESS

To a pattern scientist, the common currency of biology is not organisms, cells, or DNA—it is complex determinate patterns. In order to take advantage of the power of mathematics, the pattern biologist attempts to abstract his particular complex patterns in formal language; however, when dealing with true complexity, formal languages are necessarily incomplete representations of the real world. For this reason, it is important to see clearly the edges of these formal representations: Exactly how and where are they incomplete? The incompletenesses often lie in those places where humans obligatorily intrude.

Much of our world is created out of our own human musings.

> *To make a prairie it takes clover and one bee,—*
> *One clover, and a bee,*
> *And revery.*
> *The revery alone will do*
> *If bees are few.*
>
> *[Emily Dickinson]*

This is true even for mathematics. Sometimes it is suggested that when we finally meet an extraterrestrial race, we will communicate via mathematics, the cold, natural, and objective logic of the universe. However, cold though it may be, mathematics is not entirely separable from humans and our human emotions. Consider randomness—"random" is a mathematical concept closely allied to "complexity." "Random" lives at the farthest end of the technical spectrum of complexity: with mathematical precision we can confidently assert that many extremely complex patterns—the maximally complex patterns—are random. But mathematical as our defintion is, the meaning behind it remains baldly human.

Their answers, vague
And all at random, fabulous and dark

[Cowper]

I

"[Formerly,] the sun came up about as often as it went down, in the long run, and a coin showed heads about as often as it showed tails," explains Guildenstern in Tom Stoppard's play ***Rosencrantz and Guildenstern are Dead.*** "Then a messenger arrived. We had been sent for. Nothing else happened. Ninety-two coins spun consecutively have come down heads ninety-two consecutive times...and for the last three minutes on the wind of a windless day I have heard the sound of drums and flute..." [Stoppard, p.18]

Each morning, we exchange bedclothes for neat vestments that order and control our daily affairs and fates. Inexplicably, Rosencrantz and Guildenstern have slipped away from this predictable world. First, Rosencrantz disappears—he is killed—and Guildenstern is left alone; then, Guildenstern too is executed. And why? It is only a tragedian's caprice. The uncontrolled whims of the Fates have set the two characters naked in the universe, where the slightly foolish random falls of a coin are turned topsy-turvy and the true relentless omnipotent and cold randomness of the universe thunders and cracks forth.

Randomness—await behind the wall at the edge of an old universe, it also creeps nightly into our dreams of ancient lives. In the morning light, however, today's "randomness" maintains the mild demeanor of a proper mathematical gentleman. Have we not tamed its primeval incivility? Has the concept lost its ancient fearsomeness?

II

Humans are sensual (especially visual) creatures, and looking back into the countrysides of our great-great grandparents, we find our words tied to their lives, touching their hands, smelling their air, seeing their trees and grass and dust and rivers. "Random" is an old and wilsome word. Coming to us as *randon* through the French *randir* (to gallop), it once meant hurried, pitched with great force, impetuous, wayward, and uncontrolled. *Randir* is from the German *rand,* rim—often the rim of a shield; in Old English, "rand" may have evoked the brim of a swift river, for *Golagros and Gawane, the knightly tale* of 1470 includes: "Apone that riche river, randonit full evin, The side-wallis war set, sad to the see"—full waters rolling along the river's brink. But shield or river, the old term held at heart a bit of our special human fear of and fascination with the headstrong and uncontrolled in nature. In 1450, *Merlin or the early history of king Arthur, a prose romance* told how the horripilant dragon "caste oute of his throte so grete raundon of fiere in-to the aire...that it seemed all red."

By the end of the sixteenth century, "random" had moved to the towns. There, it took on the more mundane tones of daily commerce and conveyed mainly a sense of the haphazard and the poorly thought-out; Valentine, one of Shakespeare's Two Gentlemen of Verona, says in the play of the same name:

Valentine: As you enjoin'd me, I have writ your letter
Unto the secret nameless friend of yours;
Which I was much unwilling to proceed in,
But for my duty to your ladyship.

Silvia: I thank you, gentle servant: 'tis very clerkly done.

Valentine: Now trust me, madam, it came hardly off;
For, being ignorant to whom it goes,
I writ at random, very doubtfully.

And, by the next century, the fearsomeness of "random" could be turned unperturbedly on its head, when Sir Walter Scott wrote:

Oh, many a shaft at random sent
Finds mark the archer little meant!
And many a word, at random spoken,
May soothe or wound a heart that's broken!

As the edges of old "random" began to soften, the word also slipped into the more literary scientific descriptions. In the 1767 edition of the *Philosophical Transactions of the Royal Society of London,* the Reverend John Michell, English astronomer and geologist and the founder of seismology, propounded a mathematical demonstration of the orderly arrangement of the cosmos. His argument proceeded by overthrowing a logical straw man: Michell disproved the null hypothesis that the stars had been arranged haphazardly, that they had been "scattered by chance" or, as he reiterated with a touch of color, that they had been "scattered at random."

Arguments (such as Michell's) about the likelihood of chance events fall within the field of Probability Theory, where "random" makes its home today. As a discipline, Probability Theory was nucleated by a set of letters between the French mathematicians Blaise Pascal and Pierre de Fermat, in the summer of 1654. The Pascal-Fermat letters were concerned with monetary transactions, with quantifying certain wagering situations. However, money is abstract mathematics incarnate—it is a direct physical opportunity to quantify, to transform the messy phenomena of the world into hard quantal units with integer values of ones or twos or tens or hundreds—and wagering has come to exemplify Probability Theory. From the beginning, Probability Theory set *chance* as the standard straw man: conditions arising in an orderly fashion were the yardstick against which to assess conditions arising in an orderly fashion. Initially, the chance conditions were referred to as "conditions arising at hazard" or "conditions of equally probable events"; neither LaPlace's famous *Essai Philosophique sur les Probabilities* (1814) nor John Stuart Mill's chapter on *chance* in his *System of Logic, Ratiocinative and Inductive* (1846) used the word "random."

Mathematics evolves by rephrasing one field's problems in the language of a disparate field. In 1854, the English mathematician George Boole rewrote Probability Theory in the precise terms of his Calculus of Logic. In *The Laws of Thought,* he first introduced the phrase "random distribution" citing the Reverend Michell's paper on the clustering of stars. In all other contexts, however, Boole followed the mathematical tradition—he used "determined by chance" or "determined by hazard" instead of "determined randomly," and he wrote "all possible constitutions of the system are equally probable" instead of "all possible constitutions of the system can appear at random." Nevertheless, with Boole's *Laws of Thought* "random" had been let into the mainstream of mathematics. The word conveyed something special, an intuitive feeling that could not be dispelled or dismissed; and by the end of the nineteenth century, "random" had rooted itself in the working vocabulary of the mathematician.

To all appearances, the mathematician's "random" seemed tame and quiet. Pierre-Simon de Laplace (of whom it was said: "he would have completed the science of the skies, had the science been capable of completion") wrote:

> the theory of probability is at bottom only common sense reduced to calculus; it makes us appreciate with exactitude that which exact minds feel by a sort of instinct without being able ofttimes to give a reason for it. It leaves no arbitrariness in the choice of opinions and sides to be taken...it gives us the surest hints which can guide us in our judgments, and...it teaches us to avoid the illusions which ofttimes confuse us.
>
> [Laplace, p.196]

and, it was in this spirit that, as "random" rose to an establishment term in mathematics, it automatically took on the respectable, subdued, and somewhat fussy trappings proper to civilized mathematics.

With its acceptance as a proper mathematical entity in the field of Probability Theory, "randomness" seemed to acquire clarity. In textbooks at the turn of ths century, black and white balls were confidently being drawn at random from urns, and "true coins" were falling regularly in random sequences. The full demarcation of "random" as a respectable mathematical concept, imbued with a precise and abstract meaning and identified for independent mathematical study, soon followed. The first widely used random number table was published in 1927; then, in his 1933 axiomatization of Probability Theory, Andrey Kolmogorov wrote: "random variables...from a mathematical point of view represent merely functions measurable with respect to P(A), while their mathematical expectations are abstract Lebesgue integrals." This definition led to textbooks filled with such sentences as: "a function defined on a sample space is called a random variable" and "consider a probability space on which *n* random variables are defined'' and ''among discrete random variables those assuming only the integral values k=0,1,2,...are of special importance.''

Ideally, mathematical words should be precise, well defined, and bland: the discipline depends on the uncolorful and the nuance-poor. Hidden intimations cannot lurk behind the concepts, edges must be without fuzz, and corners must be crisp. Therefore, the explicit tie between ''random'' and the dark forces was temporarily buried, and there was no longer the sense of foreboding that gave an ominous color to the Shakespearean verses of 300 years earlier:

> 'Hard-favour'd tyrant, ugly, meagre, lean,
> Hateful divorce of love.'—thus chides she Death,—
> 'Grim-grinning ghost, earth's worm, what dost thou mean
> To stifle beauty and to steal his breath,
> Who when he lived, his breath and beauty set
> Gloss on the rose, smell to the violet?
>
> 'If he be dead,—O no, it cannot be,
> Seeing his beauty, thou shouldst strike at it;—
> O yes, it may; thou hast no eyes to see,
> But hatefully at random does thou hit.
> Thy mark is feeble age; but thy false dart
> Mistakes that aim, and cleaves an infant's heart'.

[''Venus and Adonis'']

But, although ''random''acquired the neat geometric tailoring of mathematics, it had been set squarely over a black hole of logic, and the insubstantial and the mysterious continued to ferment underneath. As William Feller points out in his authoritative treatise on Probability Theory, the word ''random'' is still not well-defined mathematically and the essential problem can be traced to that old and irritatingly hazy notion *probability,* the pneuma of randomness. Contemporary mathematics considers ''random'' to be a synonym of ''equi-probable''—but as Kolmogorov, our century's greatest probability theoretician, wrote: ''One of the most important problems in the philosophy of the natural sciences [remains] the essence of the concept of probability itself.'' [Kolmogorov, p.8]

III

None of us lives comfortably with the indefinite or the ill defined, and mathematicians have struggled to iron a definite wrinkle-free order into every corner of their universe. "Random" is ill defined because it depends on the concept of equi-probable, and here mathematicians have tried to create clearer definitions in two ways. Some mathematicians declare "equi-probable" to be a concept so elemental and primitive that it can be understood only *intuitively*. Other mathematicians propose that "equi-probable" resides only in mass phenomena and that it can be recognized only *retrospectively*, when we can sit back and securely assess the outcome of those complex multifaceted interactions arising among cloud and crowd.

Ultimately, however, neither approach satisfies, and mathematical theorists have continued to grapple with the problem; they take out their ironing boards and try repeatedly to smooth the probability corner of the universe. The intuitive approach is not satisfying because it is not mathematics. Mathematical definitions can be translated into standard symbols, whereas relying on purely intuitive notions means using ideas that cannot be written down formally; unwritten ideas are slippery items, and two reasonable mathematicians can always disagree honestly about informal definitions. On the other hand, the retrospective approach can be formalized, but it is unsatisfying because it lacks penetration and insight: it tells us nothing about the clockwork inside the events themselves, and we can make no predictions. With retrospective mathematics, we lose the magic of divination. The ideal—the mystical goal to which we all aspire—is mastery over time and space, and this we find only in philosophically deep definitions, definitions that model the inner structures of the events and of their interactions. To wizards, alchemists, and happy mathematicians have been revealed the sinews and veins within the earth and also the auguries and portents of the future.

Mathematicians need formality, and they need depth: for the crystal through which they capture "randomness"—that is, for their formal definition—mathematicans want clarity, lucid precision, and cold clean depth. The hard crystalline precision to which they aspire requires some form of determinacy, a quality that is in direct opposition to all that probability holds sacred . Thus, to attain some measure of determinacy and of control, mathematicians must somehow wrench "randomness" from "probability," and to do this, they have turned to machine-based definitions. Machines can add substance to rhetoric in mathematics. Real machines are determinate; they speak of practical operations in the real world, and they have innards on which you can get your hands. The construction, even the narrative construction, of a machine that produces some particular thing is tantamount to fabricating a deep, determinate definition of that thing. For this reason, mathematicians have pushed toward a machine definition of "randomness."

The attempts at complete machine-based definitions of "randomness" have not been fully successful. For phenomena that occur completely predictably, machines are ready-made; in its *mechanical* incarnation, the determinate phenomenon inexorably appears in the guise of lever A inevitably clicking into slot B after rachet C has irreversibly fallen one notch. In this realm, "equi-probable" and related uncontrolled concepts cast no shadows—determinate operations happen mechanically, inescapably, and reproducibly. On the other hand, for processes that defy complete advance prediction—that is, for indeterminate phenomena like randomness—mechanistic incarnations cannot be fully complete; any machine description must be supplemented with some mysterious inner spleen, a bit of vaguery that engenders the essential probabilistic and indeterminate qualities. At some point, even in a machine definition, the mathematician finds himself saying: "This apparatus will produce one of a number of events, and the probability of event (a) is..."* It seems that randomizing machines always need a bit of soul or some equivalent organ that is arcane with an impenetrable uncertainty.

---

*For instance, the English statistician Sir Ronald Fisher wrote:

> [randomisation is a] physical experimental process [that] ensures that each variety has an equal chance of [occurance].
>
> [Fisher, p. 51]

The struggles over a precise mechanistic definition of randomness have a long and thoughtful history, and all lines of attack suggest that "randomness" inescapably includes some nonmechanizable unpredictability. Nonetheless, mathematicians would like to hem in this bit of waywardness and to restrict the unruly to a small and circumscribed domain. In the mid-1960s, Kolmogorov and Gregory Chaitin succeeded in reducing the realm of indeterminacy by distancing randomness from probability and by wedding randomness to complexity instead. "Complexity" is a determinate concept—it can be written in terms of computing machines—and Kolmogorov and Chaitin showed that patterns that are maximally complex also meet all of the statistical criteria of randomness. "Maximally complex" means "very heterogeneous," and random patterns are very heterogeneous.

Beyond heterogeneity, though, " random" still retains another quality. The random clouds inching across a porcelain sky are not what they were if you take away your eye for a moment, and those shaggy sketches of random fog above the horizon have no shape that you can put in words. They are complex and heterogeneous, certainly, but cloud shapes are also unpredictable, and it is these two qualities *together* that make for true randomness. Cloud patterns are random: they are not just maximally complex, they are also generated by stochastic processes; unlike the maximally complex patterns output by Kolmogorov-Chaitin complexity machines, clouds appear capricious.

Random is complex *and* capricious. "Complexity" has a machine definition. Can "capriciousness" also be defined mechanistically? Is there such a thing as a capricious machine? My two-year old daughter is unpredictable, and her behavior is sometimes random, but for this I offer the capriciousness of the human soul, which defies formal definition. If there is any hope for a machine definition of capriciousness, then we must try to model some other, essentially inanimate, process.

Perhaps that process is coin-flipping — coin-flipping is also complex and unpredictable, and coin-flipping seems much nearer the realm of machines. Can we fabricate a mechanistic definition of coin-flipping? To begin, exactly what is a coin toss? You take a penny from your pocket and set it on your thumb nail. Is it heads up or tails up?—that depends on how the coin sat in your pocket. Now you flick your thumb, you wave your hand upward, and a copper glint arcs over your head and down into your palm. For good measure, you flip the coin again, then you slap it onto the back of your hand. Heads or tails? If the outcome is truly random, then who can guess?

How might the coin flip be constructed as a machine phenomenon, like a computer output? A particular Heads or Tails is unpredictable: too many events and too many influences conspire together in any one coin toss for us to make an accurate decision in advance. Moreover, the coin itself represents an infinitely malleable or completely naive set of elements. Heads and Tails can each appear in any order—they have no intrinsic "knowledge" as to their roles as players on the universal stage of patterns; without the script, neither they nor we know in advance what their particular acts may be. In terms of a computer program, the coin is equivalent to the symbols available for constructing the output. Like the ones and the zeros in the binary output of a Turing Machine, Heads and Tails can be arranged in any conceivable sequence; they are intrinsically naive. Thus, one aspect of coin flips might be modelled as the binary output of a computer with completely permissive "universal assembly laws" (intrinsic constraints) dictating the final sequence of ones and zeros.

Beyond intrinsic constraints, any pattern generating system can also use extrinsic guides, curfews, channels, curbs, and sanctions. Collectively and in general, the extrinsic forces are known as the templet:

> A *templet* is the set of constraints and opportunities posed by the specific situation—the real world context—in which any pattern is generated.

The templet is like a computer's instructions, setting a particular organization among the milling throng of output symbols. Templets are specific blueprints or prepatterns; they are the playscripts for the production of patterns. For coin flipping and other phenomena that fabricate complex patterns, the templets are complex, and many extrinsic forces impinge upon the intrinsic potentials to determine the final outcome. However, unlike complex computer programs, the coin-flipping templets are not available to us.

A maximally complex pattern is built of completely naive elements, and both coin-flipping and computers can produce maximally complex patterns. Coin flipping (and children and cloud-covered skies) can also produce *random* patterns, but, although computer-generated patterns can be maximally complex, they are never truly random because we can predict them in advance; the instructions—the program or templet—in a reliable computer completely determines its output and, maximally complex as that output may be, it is available to us; it is predictable and not random. Here, then, is the essence of randomness: for random patterns the specific templets are unavailable to *us* in advance. If we possess the templet, the blueprint, the instructions, or the program, then we can make an accurate prediction; if we do not have the templet, then prognostication is guesswork.

IV

Like all patterns, a random pattern takes form as a templet molds the manifold behaviors of a set of pattern elements: a detailed playscript guides the innocent actors. Ideal mathematical definitions of "random" try for full determinacy, including determinate templets. But if *we* do not know precisely the operant templets (because they are unavailable to us for random phenomena), then to ensure determinacy, mathematicians must have the templets generated entirely by machines; in particular, a parent computer must always formulate the templet programs by which the child computer would then output random patterns. When determinate machines create the templets, then even unknown templets can also be determinate.

Unfortunately, such machine-only schemes are incomplete because they lead to a beginning-less ancestral tree of templets. We can never become privy to any of the ancestral templets, or the end process will cease to be unpredictable and thereby truly random. Instead, preexisting templets must forever beget descendant templets, and looking for an ultimate beginning, we would lose ourselves in the primeval history of templets. If we are ever to rest at peace under some grassy mound, then we shall be forced to say either: "The first templet has been lost in the mists of time" or "There was no beginning, no original templet." There is no escape from such undefined beginnings—if we try for a fully mechanized description, we build on an ephemeral foundation; so, it is best that we simply admit from the start that we must live with black holes in our universe of abstractions.

For randomness, we can explicitly face this incompleteness by proposing a mixed definition at the outset. Rather than using a machine-only definition, we forge a machine-with-person definition in which the observer takes his rightful place beside the observed:

> Random patterns are those maximally complex patterns for which *we* do not happen to possess the appropriate templets in advance.

It may be, as in the case of flipping a coin, that it is impractical for us to make a definite prediction, because too many events and too many influences contribute in the creation of the templet. It may be, as in the case of the behavior of a child, that the templet is built with characteristically human capriciousness. Or it may be, as in the case of subatomic particles, that the underlying machinery is forever beyond the resolution of human science. "Unpredictability" is always synonymous with *our* lack of complete foreknowledge of the operant templet, and any definition of "random" must explicitly take into account the particular *us* of interest. Certain observers have not been shown the playscript before the drama unfolds; to them, the play is unpredictable—and if the action is complex, the play is random.

This returns us to the roots of "random," for we and our limitations have always been there in "randomness." In Shakespeare's *King Henry the Sixth, Part I,* Margaret, unable to understand completely the mumbling of the Earl of Suffolk as he talks to himself, wonders: "He talks at random; sure the man is mad." The earl is in love with Margaret, and his comments are clear and sensible to the audience; but they are not intelligible to Margaret, and for this reason they are random *to her.* Random is in the eye of the beholder—to a newborn robot, for instance, all the world would appear to be random, each new program would be an unpredictable surprise, and the machine would continually be musing: "SQR(A^2+B^2)='HYPOTENUSE', well fancy that!" and "LET X='BLACK', I never would have thought of that."

For such a mythical sentient computer to attend to randomness, however, we must infuse the machine with emotion. The poignancy of unintelligibility and unpredictability comes from a uniquely animate anguish—it is built into the animal brain—and the definition of "random" demonstrates that we cannot banish this anguish from our animate world. The emotional content of randomness is a requisite part of even the most precise and outwardly unemotional of human abstractions. Mathematicians must retain the unfathomable in "random"; as the word originally held, Nature is at times wayward and inscrutable. This natural mystery is essential if we are to tether Probability Theory to our real universe. "Probability," wrote George Boole, "is expectation founded upon partial knowledge. A perfect acquaintance with *all* the circumstances affecting the occurrence of an event would change expectation into certainty, and leave neither room nor demand for a theory of probabilities." Probability breathes with a rhythm beyond our prescience, and it is patterned by the unpredictable templets of randomness. Today's mathematician may dream that his discipline will eventually purify itself of the subjective, but he can never completely exorcise the human soul from its creations. Peculiarly human elements permeate the mathematical foundations of science, and beneath the frippery of modern mathematics, "random" still bespeaks our essential helplessness.

V

Coin-flipping is a well-worn human paradigm for watching randomness in action. It has become a comfortable way for us to dabble in the uncontrolled in nature: the capricious turns of the coin are predictable unpredictabilities over which we smile and at which we gently rail in secure collusion with our friends and families. Coin flipping is "controlled randomness": humans decide when and where to admit this particular touch of the wild and the willful into our world. In contrast, Nature's untamed and uncontrolled randomness drives like a hurricane of Brownian motion, unpredictably buffeting and knocking us about the world. In a sunbeam one autumn morning, specks of Brownian dust may seem like little children, each busily intent on his own innocent machinations, propelled by who knows what tiny but all-consuming ideas. They are dusty bits that suddenly shine as they hit the sunlight and that soar or stop or float with unselfconscious calmness down an airy waterfall that only they can see. Calm, happy, and secure in their autonomous fights, are they playing out their private little dust-born programs, like newborns who hope for a world where they can be (as Anna Freud wrote) "laws unto themselves"? Or are they instead awash in a vast sea of incomprehensible tides and whims? If so, then these bits and specks must be terrified indeed; and if she could see their tiny faces, then any mother would certainly run to hug them, for this anguish of helplessness is the dread of motherlessness. Mothers make the world of a helpless infant secure, and they make it predictable; motherlessness is the most primal loneliness.

Randomness is raw unpredictability, and its emotional content is the primal infant loneliness. Infant loneliness appeared 200 million years ago at the time of the evolution of mammals; the pain of separation has its root in the maternal-infant bond, a hallmark of mammals. As the brain genealogist Paul MacLean has pointed out, maternal care and the special mother-infant interactive behaviors (such as the plaintive infant isolation call) are both tied, in the mammalian lineage, to the evolutionary creation of the cingulate cortex. The cingulate is unique—it is a part of the limbic system of the brain that suddenly arose in the jump from cold reptile to warm mammal. The cingulate cortex is Odysseus, journeying to the land of the dead at Ocean's founts and facing, in the shade of his mother, the final bleak Hand of Fate, a random Hand that belies the harmless epithet "whimsical" with the stone-cold finality of unpredictable regimens:

> I bit my lip,
> rising perplexed, with longing to embrace her
> and tried three times, putting my arms around her,
> But she went sifting through my hands, impalpable
> as shadows are, and wavering like a dream.
> Now this embittered all the pain I bore,
> and I cried in the darkness: 'Oh my mother,
> will you not stay, be still, here in my arms,
> may we not, in the place of Death, as well,
> hold one another, touch with love, and taste
> salt tears' relief, the twinge of welling tears?'
>
> [Fitzgerald, p. 191]

Filled with this primal loneliness, "random" remains untamed. Random we are born, random we die, and random rolls through the odyssey of our life. Randomness is the storm terror as a child lies alone in bed with rain in the trees outside. Lightning deep from the farthest reaches of the night sky floods the walls and floors on high, mystically depopulating the world, laying bare pre deluvian evils unfurled from bare caves and blank crags. Then, a thunder of towered black rock that sags and suddenly splits and falls, cracking bleak primal halls—stone monuments that lean from cloud-shrouded heights, mean falling cracking towers of ice giants angered and echoing twice—where I am alone and random grown in the nighttime swirl of an empty and motherless world.

PATTERN OPERATIONS

Chapter Five

OPERATIONAL SCIENCE

Initially, the substratum of science was philosophy. However, during the Renaissance, the natural sciences changed from using explanations by a priori deduction and theoretical decree to using explanations built from direct observation and experimentation. Real-world examples, case histories, and experimental confirmations assumed scientific primacy. For instance, in England, Francis Bacon wrote: "Our course and method [is] to deduce cause and axioms from effects and experiments"; and on the continent, Galileo said: "Science is measurement." The new emphasis on empirical and inductive processes focused science on its methods and techniques—and eventually, this led to:

> *In the mind of the average [scientist today there is] a body of more or less vividly apprehended methods of procedure for dealing with a certain class of physical situations, [and these] are accepted with confidence because by first-hand experience they have been found to yield correct results*
>
> *[Bridgman, 1936, p.134]*

Today, the substratum of science is its operations.

> A poem should be equal to:
> Not true
>
> For all the history of grief
> An empty doorway and a maple leaf
>
> For love
> the leaning grasses and two lights above the sea—
> A poem should not mean
> But be
>
> [Archibald MacLeish]

I

Today, in 1986, the U.S. National Library of Medicine classifies more than 66,000 serials; of these, only 143—ie., 0.2 per cent—are identified specifically as methods or techniques journals. Here are the proportions of basic science journals in Biology that are devoted to methods or techniques:

Anatomy	0 of 140	Marine Biology	0 of 44
Biochemistry	4 of 548	Medicine	6 of 5470
Biology	2 of 640	Microbiology	0 of 310
Biophysics	1 of 111	Molecular Biology	2 of 71
Botany	0 of 43	Neurophysiology	2 of 62
Embryology	0 of 62	Paleontology	0 of 7
Ethology	0 of 33	Pathology	1 of 285
Evolutionary Biology	0 of 23	Pharmacology	4 of 466
Genetics	0 of 97	Physiology	0 of 226
Histology	0 of 48	Zoology	0 of 83
Immunology	1 of 108		

—few journals are explicit forums for procedures. Scientific journals reflect scientific thinking and scientific behavior; they also reflect scientific aspirations, and scientists aspire to be more than artisans.

The paucity of methodological journals is consistent with the common notion that science is a collection of facts, not a collection of operations, and a typical dictionary definition is:

> *Science n.,* [ME, fr. MF, fr. L *scientia* fr. *scient-, sciens* having knowledge, fr. prop. of *scire* to know]. . .2: accumulated systematized knowledge esp. when it relates to the physical world.
>
> [Merriam-Webster]

Science, says the lexicographer, is ideas, not acts. Fowler's *Dictionary of Modern English Usage* takes an even more clear-cut stand as it contrasts science and art:

> *science & art.* Science knows, art does; a science is a body of connected facts, and art is a set of directions...
>
> [Fowler, p. 517]

But is this accurate? Does the scientist build scientific knowledge by constructing his abstract edifice with pure facts and by following the concepts where e'er they lead? If this were the case, all scientists would be conceptualists first, and Biology departments would be manned fully by researchers pursuing lines of theoretical inquiry, continually stringing together facts, and accumulating knowledge through an objective taxonomy that was independent of particular technologies.

In practice, however, biologists are artisans. Biology is more often driven by methods and techniques, and experimental design follows the natural dictates of particular operations rather than emanating from organized conceptual frameworks. The concepts follow the acts, and often they follow quite serendipitously.* Were Science purely conceptual and were it fully organized around an abstract edifice of ideas and facts, then I could not turn arbitrarily to the POSITIONS OPEN section of last week's *Science* magazine and find such technical requirements as:

---

*The Nobel Prize-winning neurophysiologist Alan Hodgkin wrote:

> The record of published papers conveys an impression of directness and planning which does not at all coincide with the actual sequence of events. The stated object of a piece of research often agrees more closely with the reason for continuing or finishing the work than it does with the idea which led to the original experiments.
>
> [Hodgkin, p. 1]

> Requirements [for this position include the use of] one or more of the contemporary technologies of recombinant DNA, gene cloning and expression, protein synthesis, hybridomas, and/or molecular structure and function.
>
> Applications are invited for three tenure-track positions [in] the following areas: (i) protein crystal structure determination; (ii) nuclear magnetic resonance studies of proteins; and (iii) macromolecular computer graphics/energy calculations
>
> The Department of Biology invites applications for two tenure-track positions...
> *Plant Molecular Biologist:* Candidates with experience in gene transfer and expression in plants are sought. Expertise in transformation systems and plant regeneration are required
> *Developmental Biologist:* Candidates using molecular approaches to the study of eukaryotic development are sought. Preference will be given to applicants with expertise in hybridoma techniques.

or simply:

> FACULTY POSITION—*Molecular Biologist* using recombinant DNA techniques; specific research area(s) open.

In fact, more than one quarter of the academic research positions in Biology are explicitly formulated around a preferred technique or methodology;* each of these methodologies is based on a few clear, experimental role models or operational paradigms, and, in this way, biology has become organized by its operational paradigms.

---

*This proportion was estimated as follows: an arbitrary two-week window (the last two weeks in January) was examined for each of seven successive years (1980-1986). During this period, there were 283 advertisements in the journal *Science* for tenure-track biology research (not teaching-only) positions at academic institutions; of these, 76 explicitly described the desired candidate as pursuing research using specific techniques and particular methodologies.

II

Science is action, and scientists are practioners: "Experiment is the sole source of truth," wrote Henri Poincare. "[Experiment] alone can teach us something new; it alone can give us certainty." [Poincare, p.140] And the physicist J. Robert Oppenheimer reflected that pervading all of science "there is an element of action inseparable from understanding." [Oppenheimer, p.273] Moreover, science is not only active, it is restless—"science is not a quest for certainty; it is rather a quest which is successful only to the degree that it is continuous." [Conant, p.26]

Science, then, is unceasing operations. In "Ars poetica," Archibald MacLeish tells us that a good poem cannot proclaim, it must demonstrate, and so too with science. Good science cannot simply, dispassionately, and passively announce—it must actively enact, perform, and demonstrate. In science, even the hardest and coldest facts are basically enactments: things are ultimately defined as the stereotyped products of particular operations. Insulin, for instance—before Frederick Sanger's Nobel Prize-winning work, insulin was defined operationally; in 1944, insulin was "a clear, colorless aqueous extract of the islands of Langerhans of the pancreas." [Dorland, p. 736] This definition stemmed from the original papers of Frederick Banting and Charles Best, where insulin was defined by an extraction procedure.*

---

*For example, from Banting and Best's 1922 paper:

> The entire pancreas immediately after removal [from a dog] was cut into small pieces which were put into 0.2 per cent alcohol and allowed to stand [for two days]. It was then macerated, filtered, and the clear filtrate evaporated to dryness in a warm air current. On [the next day] this dry resin-like residue was emulsified in 25 cc. ringer's solution and 6cc. were given intravenously [to a dog].

After Sanger's work, the definition of insulin changed—it became more chemical; for instance:

> Human insulin is that polypeptide hormone with the molecular structure—
>
> *A chain:*
>
> Gly-Ile-Val-Glu-Gln-Cys-Cys-Thr-Ser-Ile-Cys-Ser-Leu-Tyr-Gln-Leu-Glu-Asn-Tyr-Cys-Asn
>
> *B chain:*
>
> Phe-Val-Asn-Gln-His-Leu-Cys-Gly-Ser-His-Leu-Val-Glu-Ala-Leu-Tyr-Leu -Val-Cys-Gly-Glu-Arg-Gly-Phe-Phe-Tyr-Thr-Pro-Lys-Thr
>
> [Windholz et al., p. 4866]

Is this chemical definition less operational than the older, extraction definition? The chemical definition describes insulin as a chain of amino acids—amino acids such as *alanine.* And what exactly is alanine? We turn to a chemical textbook. As one of the descriptions of alanine, *Organic Chemistry* by Morrison and Boyd includes an operational definition. Alanine is that amino acid which:

> can be produced by amination of an acid, specifically the direct ammonolysis of an alpha-bromo acid [through the reaction:]

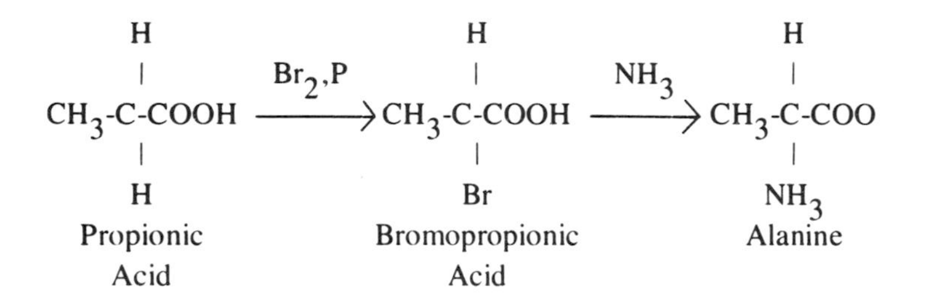

[Morrison and Boyd, p. 1123]

In addition, Morrison and Boyd offer the definition:

Alanine is an alpha-amino carboxylic acid of the structure

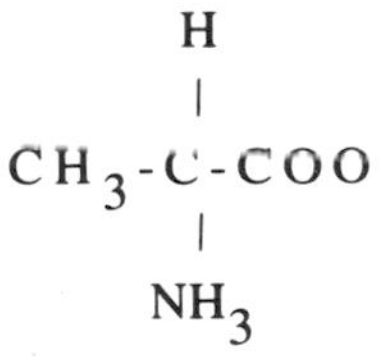

$$\begin{array}{c} H \\ | \\ CH_3\text{-}C\text{-}COO \\ | \\ NH_3 \end{array}$$

[Morrison and Boyd, p. 1118]

This is a physical definition; here, alanine is reduced to a list of atoms and atomic bonds. Is this definition less operational? Now we turn to a physics textbook and search out "hydrogen": hydrogen is defined as an atom consisting of one electron and a nucleus with charge +1e. But can we understand these subatomic particles apart from operational definitions?

Essentially, hydrogen is an electron and a proton; the electron is probably the most well understood of subatomic particles:

> Physicists have concentrated on the electron because it is the [subatomic] particle that can be most easily investigated in the laboratory. It seems to be consistent with our thinking to regard the electron as an elementary particle. No sign of an electronic substructure has been found to date.
>
> [Fritzsch, p.177]

Our current understanding of electrons is captured in their electrodynamic definition, in which an electron

> is a "cloud" of negative charge. The size and shape of the electron cloud can be calculated for a given state of an atom. For the ground state in the hydrogen atom, the [wave function—the density or amplitude of the "matter wave" at any distance r from the center of the atom with a nuclear charge

Z-] is:

$$\psi(r)=sqr(Z^3/(\pi r(o)^3)) \exp(-Zr/r(o))$$

(where r(o) is a constant called the "Bohr radius").

[Giancoli, p.802]

In other words, the electron is defined in terms of a distance and a charge. Had we lived before Einstein, these concepts might have an independent, objective, and absolute meaning; but today, elemental qualities take on meaning only in a well-defined operational context in which the measuring devices have been explicitly described.* To make any full sense of the concepts of physics and to comprehend truly our definitions, they must all be written operationally—and this means in human terms. Bridgman [p. 122] wrote:

> Any precise description of what we do in making [even] the measurements by which the various concepts of wave mechanics acquire meaning has involved a reduction of everything to classical terms. For example, the apparatus is eventually to be made

---

*As Eddington wrote:

> The vocabulary of the physicist comprises a number of words such as length, angle, velocity, force, potential, current, etc. which we call "physical quantities". It is now recognized as essential that these should be *defined* according to the way in which we actually recognise them when confronted with them, and not according to the metaphysical significance which we may have anticipated for them. In the old textbooks mass was defined as "quantity of matter"; but when it came to an actual determination of mass, an experimental method was prescribed which had no bearing on this definition...[Today,] Eintein's theory...insists that each physical quantity should be defined [solely] as the result of certain operations of measurement and calculation.

[Eddington, p. 254-255]

> heavy so that we can get up to the scale of ordinary experience. Our meanings are thus ultimately to be sought in the realm of classical [everday human] experience.

He then summarized the situation;

> In general, we mean by any concept nothing more than a set of operations; *the concept is synonymous with the corresponding set of operations.*

Science is operational—in science, things are what we do.

III

The substance of science is operations, not facts—and the goal of science is prediction, not discovery. Poincare wrote:

> Most important of all, the man of science must exhibit foresight. Carlyle has written somewhere something after this fashion. "Nothing but facts are of importance. John Lackland passed by here. Here is something that is admirable. Here is a reality for which I would give all the theories in the world." Carlyle was a compatriot of Bacon, and, like him, he wished to proclaim his worship of *the God of Things as they are.*
>
> Bacon would not have said that. That is the language of the historian. The physicist would most likely have said: "John Lackland passed by here. It is all the same to me, for he will not pass this way again."
>
> [Poincare, p.141]

Science is prediction, and science is also operational; what, then, is prediction in terms of scientific activities? Operationally, prediction is re-creation—it is repetition, repetition, repetition. Repetition is the basis for the confidence that we have in our results, in our conclusions, and in our generalities, and science is grounded in repetition. We can see this, for example, in the elemental requirements for published scientific work—the *Materials and Methods* section of a scientific paper must foster repeatability:

From the *Journal of Biological Chemistry*—

> The description of the *Experimental Procedures* should be brief, but adequate for repetition of the work by a qualified operator.

From the *American Chemical Society Style Manual*—

> Give enough detail about your materials and methods so that other experienced workers could repeat your work and obtain comparable results.

From O'Connor and Woodford's *Writing Scientific Papers In English* [p.22]—

> Provide enough details for an experienced investigator to repeat the experiments.

From Katz's *Elements of the Scientific Paper* [p.31]—

> [The Materials and Methods] are the stuff and the ways of science, and for this section your goal should be to write a description that is detailed and complete enough for any researcher to follow your directions and to repeat your observations successfully.

Repetition is the operational underlay of science, and Einstein even proposed that repetition (specifically, the repetition of a mental image) is the operational underlay of all rational thought—he wrote:

> What, precisely, is "thinking"? When, at the reception of sense-impressions, memory-pictures emerge, this is not yet "thinking". And when such

> pictures form series, each member of which calls forth another, this too is not yet "thinking". When, however, a certain picture turns up in many such series, then—precisely through such return—it becomes an ordering element for such series, in that it connects series which in themselves are unconnected.

Science is operational, and science aspires to prediction; thus, the essential scientific operations must be repeatable—in science, things are what we do and what we do and what we do again.

Chapter Six

DECOMPOSITION-WITH-BOOKKEEPING

Pattern Biology separates out three broad analytical forms. The first of these is definition. *Definition is an immediate and architectural analysis; it is the methodology of Anatomy and of Analytical Biochemistry, and one of the important definitional paradigms is* decomposition-with-bookkeeping.

I

That rock melted and recrystallized, metals rusting in the field, and water thickening with dissolved salts and suspended particles are the same transactions underlying life; that the makings and the breakings of the stones of the universe are identical with the spontaneous knitting together, the autonomous reworking, and the ultimate dissolution of cells; that the same sterile chemistry carried out in the laboratory builds the intimate breath of an organism—these are the insights of the biochemist, and they predate Western science. Classical Greek philosophers proposed that all things—living and nonliving—are composed of the same basic elements; in the fifth century BC, Empedocles regimented these elements as *fire, air, water,* and *earth.**

Fire is the dynamic element, and Plato pointed out that the dynamic body warmth must be generated by the same heating principle (the element fire) that is extant throughout the inorganic world. Plato's student Aristotle wrote in his *Parva Naturalia* that the internal heat of the body operates like a flame with our food as its fuel; and,

---

*To animate the inanimate, a special vital force—the *soul*—also joined the elemental forum, but most life processes were considered to be directly comparable to autonomous transactions of the inanimate world.

later, the Roman physician Galen said that body heat is generated as in a lamp, where the heart acts as a wick, the blood as oil, and the air we breathe as the necessary combustible. When the medieval canons were challenged in the Renaissance, natural philosophers such as Leonardo da Vinci, Rene Descartes, and Robert Hooke reiterated that the warmth of life radiated from a flamelike mechanism; and Antoine Lavoisier finally codified this idea in the late eighteen century. Lavoisier experimentally demonstrated both that oxygen is a combustible material in animal metabolism and that animals need oxygen to live. Together with Armand Seguin, Lavoisier wrote:

> Respiration is a true combustion, brought about by the combination of the carbon and of the hydrogen contained in the blood with the oxygen of the atmospheric air, a combustion entirely similar to that of a burning candle.
>
> [Partington, vol. 3, p. 473]

The concept that animal heat is the same fire as a candle flame is an ancient and biochemical insight, but it is not Biochemistry. Biochemistry is a collection of operational methods that are hidden behind such summary statements as: "D-glucose is enzymatically phosphorylated by ATP and ultimately cleaved to yield two molecules of D-glyceraldehyde 3-phosphate" and "glycogen phosphorylase, which catalyzes conversion of glycogen to glucose 1-phosphate, is a regulatory enzyme existing in active (phosphorylase *a*) and less active (phosphorylase *b*) forms" [Lehninger, p.439]; the biochemical operations give tangible and substantive meaning to the words "D-glucose," "enzymatically phosphorylated," "ATP," "glycogen," "regulatory enzyme" and "active form." A biochemist is a laboratory scientist, and today's Biochemistry has clear paradigms for its laboratory work. Originally, however, laboratory science was a jumble of techniques, and, in particular, synthetic and analytic procedures were interwoven in all of the standard chemical operations.

Biochemistry is rooted in alchemy, which sought to create and to transmute the elements, to liberate inherent forces from matter, and to generate new forces and new substances by combining the old ones. The alchemist hunched himself in his laboratory, mixing and heating powders and potions, always with an eye toward synergy and holism rather than toward an analytic understanding of nature. Dissection, dissociation, dissolution, and extraction were subordinated to the grand goals of making new compounds and of harnessing the generative forces of complex matter. Where philosophers and scientists such as Paracelus (1493-1541) and even Newton* persisted in traditional alchemy, they remained off the main ontogeny of Biochemistry, and they contributed little to building the modern chemical sciences. Instead, the foundations of Biochemistry came from Jan Baptista van Helmont, Robert Boyle, Robert Hooke, Joseph Black, Henry Cavendish, Joseph Priestly, and especially Antoine Lavoisier, who framed specifically analytical paradigms. These scientists distilled the analytic method from the hodgepodge of extant alchemical techniques, and their work was characterized by the methodical dissociation of compounds and by the careful quantitative tabulation of the constituent units—the fundamental steps in the decompositon-with-bookkeeping paradigm of contemporary Biochemistry.

---

*Isaac Newton

> very rarely went to bed before two or three of the clock, sometimes not till five or six...especially at spring and fall of the leaf, at which time he used to employ about six weeks in his elaboratory, the fire scarcely going out either night or day, he sitting up one night and I another, till he had finished his chemical experiments, in the performance of which he was the most accurate, strict, exact,...the elaboratory...was well furnished with chymical materials as bodyes, receivers, heads, crucibles, &c... which was made very little use of, y[e] crucibles excepted, in which he fused his metals; he would sometimes, tho' very seldom, look into an old mouldy book w[ch] lay in his elaboratory, I think it was titled *Agricola de Metallis*, the transmuting of metals being his chief design, for which purpose antimony was a great ingredient.
>
> [Newton's lab assistant (1728), quoted in Partington, vol. 2, p. 471]

II

The complexities of Nature's luxuriance proffer an unending variety of particular cases; thus, specific scientific questions are endless. At base, however, the myriad specific questions are all children of three general questions: given some object of study—some pattern—we would like to know—

1. What is the pattern?
2. How did it come about?
3. Where does it go and what does it do?

These three parent questions are the questions of the present, the past, and the future of a pattern; they are the questions of definition, explanation, and action, and of form, formation, and function. The first of these general questions is the "What" question, and the analytic operation *decomposition-with-bookkeeping* exposes the full internal architecture of a pattern—it is the paradigm of careful dissection. Specifically, it names and accounts for all the constitutent parts of a pattern, and then it proposes a complete and detailed model recreating the whole.

III

For Biochemistry, the decomposition-with-bookkeeping paradigm developed slowly—technical problems made it difficult to carry out. The first problem was *bulk purity*: starting materials must be pure and available in quantities sufficient for a complete analysis. Inorganic salts and metals could often be mined and purified, but a persistent bane of the biochemist has always been that most organic compounds occur in small quantities and are invariably mixed in among other organic compounds. The second problem was *controlled fracturing:* biochemists need to break their tiny compounds along natural planes, and they must cut gently enough to avoid disintegrating the elemental units themselves but vigorously enough to separate these units completely. The third problem was *bookkeeping:* the analyst must collect, recognize, and tabulate all of the dissected units. Technically, these three problems were overcome with microscopic separation techniques (such as chromatography, electrophoresis, and the orders-of-magnitude amplification techniques like molecular cloning), with more discriminating biochemical scalpels (progressing from fire through acid to selective enzymes), and with selective, controlled, and effulgent labelling techniques (such as radioactive and fluorescent molecular tagging methods).

Such technical advances mold the conceptual base of a discipline; researchers set out the range and the form of a science as much by their experiments as by their explanatory writings. In fact, each field of study is grounded in a set of archetypic experiments—operational paradigms—and a paradigm is usually associated with a particular researcher. In the first half of the nineteenth century, Justus von Liebig produced the first elementary analysis of lactic acid, he discovered the amino acid tyrosine, and, both by precept and by example, he outlined the compositional analysis of proteins.

Next, Emil Fischer set the operational role models. In an orderly series of breaking-and-bonding experiments, Fischer unpuzzled the structures of a great many organic molecules, including the rosaniline dyes, various sugars (notably: the aldohexoses, the ketohexoses, the disaccharides, and certain glycosides), and members of the xanthine family, like caffeine. He organized the architectonics of the study of proteins, showing that proteins are linear strings of amino acids; he inferred the universal importance of the peptide bond in proteins, he coined the term "polypeptide," and he fabricated a variety of simple polypeptides by condensing amino acids in the laboratory. He also synthesized the amino acid serine, and he was the first chemist to synthesize a nucleic acid.

After Fisher, in the first half of this century, protein biochemistry was still in its "Linnaean" days—proteins were classified by analogy, not by homology. Proteins were distinguished by their size and weight, they were categorized by shape (*fibrous* or *globular*), they were grouped by general composition (*simple*—composed solely of amino acids, *conjugated*—also composed of other molecules such as sugars, or *derived*—generally laboratory breakdown products of natural proteins), and they were divided into *water-soluble* (albumins, histones, and protoamines) and *water-insoluble* (albuminoids, globulins, and prolamines).

Analogies are similarities without genealogy: they are based on common characteristics to which we have chosen to attend without regard to ancestral relations, and tabulations of analogies are ad hoc.

In contrast, homologies are inherited from common ancestors: family trees are classification by homology. In the 1930s and 1940s, biochemists undertook the analogistic task of understanding the physical basis of protein functioning; but, by the 1960s, biochemists were immersed in homologistic classification schemes because a forest of protein family trees had grown up around them.

As they began to unveil the internal architectures of complex molecules, biochemists realized two things: the functioning of a protein depends on its geometry, and the geometry of a protein is a direct consequence of its amino acid sequence. For these reasons, biochemists now took on the challenge of deciphering the micro-molecular topology within protein molecules—and this meant sequencing individual proteins:

> It is the hope and expectation of the protein chemist that the chemical, physical, and biological properties of each protein may be explained in terms of its exact structure,

wrote White, Handler, Smith, and Stetten, Jr. in the first edition of their textbook *Principles of Biochemistry* [p.169]. In 1954, when this was written, the exact structure was known for only one natural protein: the hormone insulin had just been fully sequenced by Frederick Sanger and his colleagues. Why insulin? Why was this the first protein to be sequenced? Professor Sanger writes:

> With regard to the insulin work, this was essentially a continuation of the work of A.C. Chibnall and his colleagues. When I had completed my Ph.D. in 1943 Chibnall came to Cambridge as the new Professor of Biochemistry and offered me a position in his group. They had been working on protein structure, particularly amino acid analysis, and had done a good deal on insulin—which was chosen because it was one of the few proteins that could be bought in a pure form, and because of its medical interest. It was in fact a lucky choice because it was a small molecule, but this was not known at the time.
>
> Professor Chibnall's group had analysed the free amino groups by the van Slyke method and found

> that there were more free amino groups than could be accounted for by the lysine content, thus suggesting that the molecule was built up of short polypeptide chains. He suggested that I should try to identify these amino groups to find out which residues they were located on, and this is in essence how I started working on insulin.
>
> The work went successfully with the development of the DNP technique, and the various other findings on insulin developed from this. The work was certainly not started with the idea of working out the complete sequence of insulin. Like most research, the next experiment was conditioned by the success or otherwise of the experiments that had gone before.
>
> [Sanger, personal communication]

Sanger's methodical sequencing of insulin set the standard paradigm for protein sequencing in general. Detailed protein sequences were the first step in the "post-Linnaean" exercise of categorizing proteins by homology rather than by analogy; protein sequences then led to DNA sequences, and DNA sequences can be used to infer those historical relationships that form the bases of natural molecular homologies.

IV

The decomposition-with-bookkeeping paradigm is internal architectural analysis, inferring the invisible composition and the inner structure of an item initially seen only globally. The first step in this paradigm is to secure pure objects of study, a working collection containing things with a particular definite composition and a particular definite architecture. When seen as patterns, pure objects of study have:

a. one and only one list of constituent elements, and an element either definitely is or definitely is not included in this list;

b. one and only one list of interconnections, in which

any pair of elements either definitely is or definitely is not linked.

In the decade between 1945 and 1955, Frederick Sanger determined the first complete and exact sequence for a natural protein, insulin—and he began his work with pure insulin. To purify a set of objects, one must first cull them from the surrounding objects and pick off the bits of weeds in which they may be entangled. For insulin, the practical operations begin with the collection of pancreases (usually from cows, pigs, or sheep); insulin is manufactured in the beta cells of the Islets of Langerhans, where it remains highly concentrated, and keeping the pancreas and discarding the rest of the animal results in an immediate one hundred thousand-fold purification. Next, the pancreatic cells are broken open by homogenization, large particles are filtered out, the proteins are precipitated with a mild acid solution, lipids are dissolved with an ether-ethanol solvent, and the precipitated protein powder is then dried.

Is this insulin powder sufficiently pure? To assay purity, we must stain our collection of objects, and for biology it is best to use Nature's stains. There is no single "right" way to split the world into elemental units—natural objects are not imbued with absolute and fundamental characteristics. The appropriate units are determined by our frame of reference; and within a defined frame of reference, we should ask Nature what she considers to be natural units. As the biochemist Konrad Bloch pointed out:

> The ways chosen by Nature for making organic compounds are easy enough to rationalize—by hindsight. Biochemistry remains, by and large, an empirical science and predicting the [content and configuration] of even the simplest biochemical reactions, on whatever grounds, remains a venture of risk.
>
> [Bloch, p.145]

For insulin, one natural assay is the extent to which the compound lowers blood sugar, and this hypoglycemia test, applied to the grayish powder that we have obtained from pancreatic homogenates, would give an activity of about 50 percent of the accepted international standard for pure insulin crystals.

Our goal is to make definite statements about the architecture of insulin molecules. Eventually, we would like to say that each molecule has exactly three alanines, and we would like to say this with complete assurance—with 100 percent accuracy, not with 50 percent accuracy. The incondensable complexities of the real world conspire to make 100 percent accuracy a dream; shades and gradations are the true substance of the complex natural world. To lay an artificial but necessary black-and-white definition on our rich, textured, and colorful universe, we set thresholds and we lay down lines above which we arbitrarily consider things to be 100 percent pure. Customarily, scientists choose a threshold of from 95 to 99 per cent for many assays—and we can reach this level for our insulin powder by specifically crystallizing out the insulin molecules.* It is this crystallized insulin that formed the starting material for Sanger's sequencing work.

First we purify, and, after purification, *decomposition-with-bookkeeping* methodically unpuzzles the content and the configuration of a pattern in two broad operations: (a) the pattern is dissolved into its constituent unit elements, and (b) the natural linkages between the elements are identified. The content is a list of all the constituent elements of a given pattern, and this list need have no particular order (technically, it is an arbitrarily-ordered set); to determine its content, we can simply dissolve the pure objects and collect the bits and pieces. Here our only bookkeeping requirement is that we account for all the pieces: for content, we need conservation of mass, not conservation of order. Therefore, the central operations of *decomposition-with-bookkeeping* begin with dissociation. But how disruptive should we be? Again, we define our frame of reference, and then, within the specified arena, we ask Nature. In terms of insulin, Nature builds

---

*John Abel (1857-1938) developed the first techniques for crystallizing pure insulin, invented the artificial kidney, pioneered standard blood transfusion methods, and founded the *Journal of Experimental Medicine* (1896), the *Journal of Biological Chemistry* (1903) and the *Journal of Pharmacology and Experimental Therapeutics* (1909).

proteins by stitching together amino acids on tiny ribosomal sewing machines; thus, for the protein biochemist, amino acids and strings of amino acids (polypeptide chains) are natural units.

Frederick Sanger began his definition of insulin by discovering its content. First, he dissociated the whole protein into its constituent polypeptide chains. Each insulin molecule is built of two side-by-side polypeptide chains (Figure 6.1) linked together by disulfide bonds between the cysteine amino acids; Sanger cut the links between the two polypeptide chains (A and B chains) with performic acid, leaving cysteic acid groups in place of the bonds. Sanger had now transformed his working set of objects from a homogeneous population into a heterogeneous population—once again a mixed collection of molecules (half A chains and half B chains) sat upon his laboratory bench. But Sanger needed a homogenous set of molecules: after dissolving your pattern, you must repurify. Each of the insulin polypeptides will preferentially precipitate at a different pH, a different salt concentration, and a different alcohol concentration. Therefore, Sanger separated out the two chains by differential precipitation, and then he quantitatively stained to assay the purity; specifically, using the differences in the amino acids at the N-terminal ends, Sanger found that the A chains had a purity of 99 percent and that the B chains had a purity of 97 percent.

With pure compounds again, dissolution is the next operation. Acid solutions break peptide (amino acid-amino acid) bonds in proteins, so Sanger now dissolved his polypeptide chains in acid (5.7N hydrochloric acid at 100°C for two days), producing a collection of individual amino acids. Again this complicated his working preparation: a pure collection of objects of study (B chains*) had been transformed into a heterogeneous collection of objects of study (various amino acids). This time, Sanger purified using paper

---

*Sanger worked simultaneously on defining both of the insulin polypeptides, but for technical reasons the structure of the B chains was unpuzzled first.

	A chain	B chain
N-Terminal ends	Gly	Phe
	Ile	Val
	Val	Asn
	Glu	Gln
	Gln	His
	Cys	Leu
	Cys ———	Cys
	Ala	Gly
	Ser	Ser
	Val	His
	Cys	Leu
	Ser	Val
	Leu	Glu
	Tyr	Ala
	Gln	Leu
	Leu	Tyr
	Glu	Leu
	Asn	Val
	Tyr	Cys
	Cys	Gly
	Asn	Glu
		Arg
		Gly
		Phe
		Phe
		Tyr
		Thr
		Pro
		Lys
C-Terminal ends		Ala

Figure 6.1
The amino acid sequence of bovine insulin.

chromatography,* and, from the chromatographic results, he found the full content of the B chain of bovine insulin to be:

no./chain	*amino acid*	
2	alanine	(Ala)
1	arganine	(Arg)
1	asparagine	(Asn)
2	cysteine	(Cys)
2	glutamic acid	(Glu)
1	glutamine	(Gln)
3	glycine	(Gly)
2	hystidine	(His)
4	leucine	(Leu)
1	lysine	(Lys)
3	phenylalanine	(Phe)
1	proline	(Pro)
1	serine	(Ser)
1	threonine	(Thr)
2	tyrosine	(Tyr)
3	valine	(Val)

Next, Sanger set about discovering the configuration of insulin, for "just as when discussing a house it is the whole figure and form of the house which concerns us, not merely the bricks and mortar

*In chromatography, a mixture of molecules in an appropriate solvent (such as butanol-acetic acid) is drawn up into a sheet of wet filter paper by capillary attraction; different molecules will move at different speeds, and Sanger's animo acids eventually formed discrete layers within the chromatographic paper. To immobilize the amino acids, Sanger dried the filter paper; and, to make the layers of amino acids visible, he sprayed the paper with a ninhydrin solution. He then decoded the stain pattern by comparing it to a comparable filter paper through which a known set of animo acids had been run.

and timber, so in Natural Science it is the composite thing, the thing as a whole, which primarily concerns us.'' [*Aristotle,* 1937] Of all natural objects, biological patterns are most keenly characterized by their configurations; the constituent units of life repeatedly are found to be arranged in only a small number of all the many possible ways.

Animate configurations are unique and idiosyncratic. There is, for instance, the stereospecificity of biochemical coumpounds. Organic molecules can take on a number of different spatial geometries, depending on the absolute arrangement of the molecular groups attached to the four bonds of an asymmetric carbon atom. If there are *n* asymmetric carbon atoms, then the molecule can take on 2^n different geometries. The amino acid alanine has one asymmetric carbon atom, and alanine can take on two different geometries or stereoisomers:

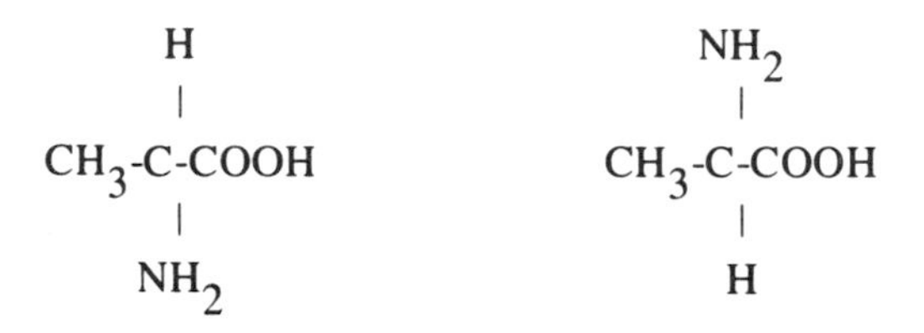

Stereoisomers have identical general chemical properties—the same melting points, solubilities, dissociation constants, and densities—but organisms are usually interested in only one of the possible configurations (most often the L form) of these compounds, and their cellular machineries, such as receptors and enzymes, can be exquisitely tuned to one particular molecular geometry. Identical contents give many identical physical properties, but the particular natural configuration defines the biological activity; content without configuration has weight but no bite, and the keen edge of biological matter is its idiosyncratic yet stereotyped configuration.

As with the content, the configuration of a pattern can be encapsulated in a simple, arbitrarily-ordered list, and matrices are the standard forms for listing configurations. Each row of the matrix repesents a constituent element, and each slot of the matrix corresponds to a possible pairing between elements. In topological terms, two paired or linked elements are adjacent, and this gives the configuration matrix its name, the "adjacency matrix." The adjacency matrix for the first seven amino acids in the B chain of insulin is:

	Phe	Val	Asn	Gln	His	Leu	Cys
Phe	0	1	0	0	0	0	0
Val	1	0	1	0	0	0	0
Asn	0	1	0	1	0	0	0
Gln	0	0	1	0	1	0	0
His	0	0	0	1	0	1	0
Leu	0	0	0	0	1	0	1
Cys	0	0	0	0	0	1	0

In an ideal world, scientists would build adjacency matrices by directly examining each possible pair of elements to determine whether there is really a linkage between them. Had Sanger an ideal microscope, he could have examined each of the amino acids in an insulin molecule to discover to which other amino acids it was bonded; checking each possible pairing, he would have put a 1 or a 0 in the appropriate matrix slot and immediately have constructed the adjacency matrix that accurately summarizes the configuration of insulin. However, Sanger found himself in the more common situation where it was technically impossible to examine directly each pairing between the constituent elements. Instead, he inferred the configuration of the insulin molecule from indirect observations: first, he proposed an extreme model, and then he modified the model iteratively until it matched the true architecture of the molecule.

Sanger's program for indirectly unpuzzling configuration can be called "the equivalence class strategy." To begin, Sanger hypothesized that all possible links exist in his object of study and that the configuration is a completely interconnected knot. Then he pruned hypothetical links as he discovered which ones were not actually present in the real world. To carry out his plan in practice, the complete object is fractured into pieces: we hit it with a "hammer" and then collect the fragments. (Each fragment becomes an "equivalence class," because all of its elements share the common equivalent property that they are attached together within the same piece.) For each fragment, the full connectivity of edge elements remains uncertain—to fracture our object, we broke some of its interconnecting bonds, and those bonds originally linked elements that are now along the edges of the pieces. In contrast, all of the bonds of central elements are preserved. Therefore, we mark the edges of a fragment of delineate its center. Central elements cannot have been interconnected between pieces; so, we take our extreme knot model and eliminate any possible connections between the central (unmarked) elements of one piece and the elements of the other pieces.

To follow this strategy and in particular to sequence insulin, Sanger invented a new, sensitive, and widely applicable method for identifying the edges of protein fragments by specifically labelling the N-terminal ends of polypeptide chains.† With this labelling technique, Sanger could mark the edge elements of protein fragments of almost any size, and he applied this technique to fractures of oxidized insulin. For instance, the edge-labelled equivalence classes of one experiment with B chains included:
(Asn, Gln, His, Phe*, Val) (Asn, Phe*, Val) (Ala, Gly, Leu*, Val) (Ala*, Leu, Tyr) [an asterisk denotes an edge amino acid]. Sanger and his colleagues repeatedly fractured the A and the B chains of insulin using hydrolysis (acid treatments): after fracturing, they listed all of the different fragments in the form of edge-labelled equivalence classes, and finally they compared the classes. With the comparisons, Sanger took the extreme knot model of the insulin molecule and pruned hypothetical bonds between central (unmarked)

---

†To mark these edges, polypeptides are reacted with the reagent 2,4-dinitrofluorobenzene (DNFB). Under mildly alkaline conditions, this transfers a yellow tag molecule (a dinitrophenyl group) to the free amino group at the N-terminal end of each chain:

$$O_2N\text{-}\langle O \rangle\text{-}F\ (\text{with } NO_2) + H_2NCHC\text{-}NHCHC\text{-}\ (R,\ R') \longrightarrow O_2\text{-}\langle O \rangle\text{-}NHCHC\text{-}NHCHC\text{-}\ (NO_2,\ R,\ R') + HF$$

DNFB | peptide | DNP-tagged peptide

amino acids and amino acids in companion equivalence classes. For example, using the sample fragments listed above, he inferred that the amino acid asparagine (Asn) is not directly linked to the amino acid tyrosine (Tyr) in insulin B chains.

For a full configurational analysis, one needs a diverse and varied collection of pieces, and many fracturings will lead to many different fragments. In their first paper, Sanger and Hans Tuppy listed the full content of the insulin B chain, but they could not yet infer its complete configuration (Figure 6.2). Two technical problems with the "hammer" of acid hydrolysis had made it diffucult to fracture the chain into equivalence classes that were sufficiently varied to resolve all the bonds. The first problem was that certain clusters of amino acids (the polar amino acids) could not be fractionated cleanly, so the polypeptide fragments of acid hydrolysis could not always be well distinguished. The second problem was that acid hydrolysis always cleaved the peptide bonds preceding serine and threonine. These amino acids were always edges of the resulting equivalence classes; thus, Sanger and Tuppy could not infer the natural connections of serine and of threonine in the whole B chain. In their second paper, Sanger and Tuppy used a different set of "hammers" to fracture the polypeptide, breaking the molecule with the proteolytic enzymes pepsin, chymotrypsin, and trypsin. This produced a wider varity of polypeptide fragments, and Sanger was then able to resolve the full series of amino acid linkages.

To complete the equivalence class strategy, we must discover the interconnections between those elements that have frequently been along the edges of our fragments. For this, we compare all the small pieces that we have been able to isolate. If two elements are not actually attached in the whole object, then small fragments of the object will rarely contain both elements. Pairs of elements never found together in small pieces will not have been directly interconnected in the complete object, and we can eliminate hypothetical bonds between these elements. Therefore, to end his analysis, Sanger compared all of the smaller fragments . Here he found, for instance, that no short pieces of the B chain contained both serine and threonine or both asparagine and tyrosine, and he inferred that in the complete insulin molecule serine is not attached directly to threonine and asparagine is not attached directly to tyrosine.

insulin	*Sanger and Tuppy 1951a*					*Sanger and Tuppy 1951b*
Phe	Phe					Phe
Val	Val					Val
Asn	Asn					Asn
Gln	Gln					Gln
His	His					His
Leu	Leu					Leu
Cys	Cys					Cys
Gly	Gly					Gly
Ser		Ser				Ser
His		His				His
Leu		Leu				Leu
Val		Val				Val
Glu		Glu				Glu
Ala		Ala				Ala
Leu						Leu
Tyr			Tyr			Tyr
Leu			Leu			Leu
Val			Val			Val
Cys			Cys			Cys
Gly			Gly	Gly		Gly
Glu				Glu		Glu
Arg				Arg		Arg
Gly				Gly		Gly
Phe						Phe
Phe						Phe
Tyr						Tyr
Thr					Thr	Thr
Pro					Pro	Pro
Lys					Lys	Lys
Ala					Ala	Ala

Figure 6.2
Steps in Frederick Sanger's inference of the primary amino acid sequence of bovine insulin.

Finally, Sanger modelled the complete insulin molecule. Comparisons of models with observations drive us back to the world of experimentation; intermediate steps, events, and inferences have intervened between the actual object and our re-creation, and we can check and revise our model by stepping back and forth between the abstract realm of theory and the real world of observations. It is both the rich variety of nature and the inability of man to capture perfectly any complex phenomenon that forces science to be iterative. Sanger tested his model by iterative experimentation—as he wrote:

> In this work many more peptides were studied from both acid and enzymic hydrolysates than were actually necessary to deduce the sequence. This was considered essential since the methods used were new and were qualitative rather than quantitative. The fact that all the peptides fitted into the unique sequence given above added further proof of its validity.
>
> [Sanger, 1960, p. 186]

The real world is complex, but our most satisfying abstractions are simple. At the same time, no simple model completely captures nature—each inevitably smooths rough edges and turns a blind eye to variations. When set against the real phenomenon, when challenged with further experimentation, explicit models honestly highlight this disparity, because models are only guides for future science, and researchers model in order to remodel.

By 1955, Sanger and colleagues had deciphered the complete sequence of insulin—they knew the content and the configuration of both polypeptide chains, and they also knew the placement of the disulfide bonds—and this work became a dramatic demonstration of the power of the decomposition-with-bookkeeping paradigm for reconstructing the hidden architecture of a complex natural pattern. Sanger drew a detailed model of insulin, and he was then struck by its ''indecipherability'' and ''incoherence'': although the precise molecular structure—critical to the functioning of the hormone—was stereotyped, the amino acid pattern made no simple sense. Sanger [1960, p. 187] wrote:

> Examination of the sequences of the two chains reveals no evidence of the periodicity of any kind, nor does there seem to be any basic principle which determines the arrangement of the residues. They seem to be put together in an order that is random, but nevertheless unique and most significant, since on it must depend the important physiological action of the hormone.

Organisms are random patterns captured and elevated to determinate nodes in an unfolding natural sequence, and careful biological analysis had once again bumped up against the intricacies of these, the incondensably complex natural patterns of the world.

V

Descriptions of scientific paradigms should be set out in explicit unambiguous formulations, and for this there is no better structure than the algorithm—the exigencies of the computer age drive us toward precise formulaic descriptions. *Decomposition-with-bookkeeping* is an operational routine, an automatable recipe for determining the full architecture of a complex pattern. In its most abstract form, this paradigm builds two things: an arbitrarily-ordered list of the constituent elements (the content) of the pattern, and an adjacency matrix detailing the interconnections (the internal configuration) of the pattern. As an algorithm, this paradigm is:

MAIN PROGRAM

Purify

Call the *Purification* subroutine

Determine the content

Dissolve the pattern

Repurify: call the *Purification* subroutine

Repeat

List the content

Determine the configuration

Begin with an adjacency matrix completely filled with ones as a working hypothesis (the "extreme knot model")

Call the *Equivalence Class* subroutine

Pare connections from the evolving model (remove potential connections between central elements of one piece and all elements in companion pieces)

Repeat

Compare small equivalence classes to finish paring the model

List the final adjacency matrix

Build a model

Check the model and then remodel

SUBROUTINES

Purification **subroutine**

1. **Separate out a set of objects of study**
2. **Stain and assay the objects to determine their purity**
3. **Accept or Reject the purity level**
   (If P<Minimum Acceptable Standard, then return to step 1, otherwise go on)

Equivalence Class **subroutine**

1. **Fracture pattern into pieces**
2. **Mark the edge elements**

Chapter Seven

ALTERNATIVE ARCHITECTURES

A second broad analytical operation of Pattern Biology is explanation. *Explanation is a retrospective, ancestral analysis; it is the methodology of Embryology, and one if its major paradigms is* alternative architectures.

> I dwell in Possibility
> A fairer house than Prose,
> More numerous of windows,
> Superior of doors.
>
> [Emily Dickinson]

I

As they travel the path from egg to birth, embryos have an itinerary, a playscript of transmogrifications that they follow doggedly. Organisms are not created whole *ex nihilo,* and their course of development is neither random nor variable; rather, the wispy beginnings of life follow along a set plan. There exists a stereotyped embryogenic agenda—this is the embryologist's principle. Aristotle, the Peripatetic natural philosopher, was the son of a physician, and he himself practiced medicine for a time in Athens; the embryologist's principle can be traced back at least as far as Aristotle,* who argued for a pre-existing and predictable embryogenic plan:

---

*Ernst Mayr [p. 87] wrote that "no one prior to Darwin has made a greater contribution to our understanding of the living world than Aristotle"—while Darwin himself said:

> From quotations which I had seen, I had a high notion
> of Aristotle's merits, but I had not the most remote notion

> Empedocles was wrong when he said that many of the characteristics which animals have are due to some accident in the process of their formation, as when he accounts for the vertebrae of the backbone by saying "the fetus gets twisted and so the backbone is broken into pieces": he was unaware (a) that the seed which gives rise to the animal must to begin with have the appropriate specific character; and (b) that the producing agent was pre-existent: it was chronologically earlier as well as logically earlier: in other words, men are

---

> what a wonderful man he was. Linnaeus and Cuvier have been my two gods, though in very different ways, but they were mere schoolboys to old Aristotle.
>
> [Peck, frontispiece]

Strong praise all around, yet many a fine biologist has not read Aristotle.

Philosophical issues—those problems that are intimately tied to people and to our particular ways of grappling with the world—are never entirely new: the same enigmas have been debated for eras immemorial. Nevertheless, each thoughtful person must make his own sense of these issues, and this means that each researcher must, at least in part, create such problems anew; to master an issue, we must discover it ourselves and reform it in our own language, from our own unique perspective, and for our own particular times. For this reason, one cannot depend blindly upon other thinkers. Thus, it is not always intellectually honest to trace the seeds of an individual's progress far back through cultural history, even for the most fundamental, ancient, and oft discussed philosophical problems. The true intellectual history of one person's struggles with great issues may neither parallel nor be cleanly rooted in the overall history of a culture's struggles with the same issues; as one da Vinci scholar has written:

> There is great interest in relating Leonardo [da Vinci's] work to that of his medieval predecessors. However, as I delved deeper into Leonardo's own notes I found their relations to both his predecessors and successors ever more elusive and difficult to evaluate. Leonardo's thought-patterns emerge increasingly more clearly as he weaves his own personal web of ideas. He drew widely on past sources as his own library lists... But what he took he digested, absorbed and transformed into his own personal creative form of science, often rendering such gleanings almost unrecognizable.
>
> [Keele, p. 1-2]

> begotten by men, and therefore the process of the child's formation is what it is because its parent was a man.
>
> [Aristotle, 1943 (Loomis), pp. 44-45]

This idea of a set embryogenic plan was built into the anatomical tradition; it was implicit in the stereotyped descriptions of the various stages of the development of the fetus, beginning with the classic descriptive embryology of Aristotle and progressing through Galen into the Renaissance with Leonardo da Vinci, Hieronymus Fabricius, and Marcello Malpighi and then into the eighteenth century with Albrecht von Haller.

At the same time, biologists also divised more theoretical representations of the embryogenic plan; these theoretical descriptions ranged from the programmatic to the figurative. Programmatic descriptions included the Greek hebdomadic (built of "sevens") formulation in which important stages included the "critical" seventh day of pregnancy, the seventh month of gestation when the embryo is fully viable, the seventh month after birth when the first teeth appear, the seventh year of life when the milk teeth appear, and the fourteenth year of life when puberty begins. Figurative descriptions included the famous shipbuilding analogy:

> Galen compares the construction of an animal to [the building of] a ship. For, says he, as the foundation and commencement of a ship is the keel, from which the curved ribs extend on either side at moderate distances from each other like a hurdle so that the whole fabric of the vessel may afterwards be built up on the keel as a suitable foundation, so, in the animal structure, Nature employs the outstretched spine and the ribs curved out from it as if they were a keel or suitable foundation laid down for it, and then builds and completes the whole pile.
>
> [Fabricius, in Adelmann, 1967, p. 200]

Such representational descriptions of the embryological stages are definitional, and they lie in the anatomical stream of biology. In contrast, the embryologist's task is to go behind description, to trace out the history of each of these stages, and to understand the embryogenic processes: embryologists aim for explanations.

Embryological stages are biological patterns, and patterns can always be partitioned into two aspects: their elements (their content), and their topology (their configuration). Traditionally, this same fundamental split has been used to classify embryogenic processes: embryogenic processes have either been called ''transformational'' (giving rise to a particular content), or they have been called ''conformational'' (giving rise to a particular configuration). This division was explicitly set forth by Claudius Galen, logician, philosopher, and physician, ''who, above all others, gathered up into himself the divergent and scattered threads of ancient medicine, and out of whom again the greater part of modern European medicine had flowed.'' [Allbutt, p. 44] In *On the Natural Faculties,* Galen wrote that embryogenesis is ''compounded of'' two processes:

(a) ''The underlying substance from which the animal springs must be *altered*'' [i.e., transformational change].

(b) ''The substance so altered [must] acquire its appropriate shape and position, its cavities, outgrowths, attachments [through] a *shaping* or *formative* process'' [i.e. conformational change].

[quoted in Needham, p. 70]

Galen's perspective was adopted whole by such pioneering embryologists as Fabricius, and it continued to organize embryological thinking well into the eighteenth century. The transformational events in embryogenesis—the generation of apparently new content—while mysterious, were alchemical; in transformation, new elements are created from old, and with the emergence of Biochemistry, scientists began to understand how new elements, new parts, and new substances of an embryo might be produced. On the other hand, the conformational events—the generation of apparently new configurations—were the central enigmas of Embryology: more than any other biologist, the embryologist was forced to wrestle with the difficult problem of the stereotyped creation of specialized configurations from embryogenic systems in which these particular forms were not previously apparent.

Embryological configurations are the regular, complex, and concordant organizations of the embryo; they are a great wonderment,* and two schools of thought—preformationism and epigenesis—arose to explain the appearance of these special embryological configurations. Preformationists proposed that configurations are not created *ex nihilo;* instead, topology is inherited during each embryogenic transformation. Thus, Lucius Seneca (tutor of the emperor Nero) wrote—

---

*On the placement and the shape of the eyes, Fabricius marvelled:

> The formative faculty, which makes the homogeneous dissimilar, conferring upon them comeliness through a suitable form, a proper size, a fit position, and a suitable number, is...endowed with the greatest wisdom...After the eye, for example, has been engendered by the alterative faculty, the task of placing it in the head and not in the heel and of giving it a round instead of a square shape, or other form, a proper size, and a number which is neither one nor three or more, this task, I say, seems not to be accomplished naturally but rather with discrimination, knowledge, and intelligence. The formative faculty seems undoubtedly to have exact knowledge and foresight both of the future action and of the use of each part and organ, foreseeing as if indeed endowed with infinite wisdom, that eyes are provided to see with and that they will be serviceable for vision if they lie in an elevated place, so that, as from a watch tower, they may view and survey all; and foreseeing, too, that they ought to be fashioned round in shape that they may be instantly moved in every direction to see everything, and that two eyes are the suitable number so that they may see more, and so that if one is injured the other may retain at least half of their action.
>
> [Adelmann, 1967, p. 198]

> In the seed are enclosed all the parts of the body of the man that shall be formed. The infant that is borne in his mother's womb has the roots of the beard and hair that he shall wear one day. In this little mass likewise are all the lineaments of the body and all that which Posterity shall discover in him.
>
> [quoted in Needham, p. 66]

Later advances in optical and other investigative instruments did not disillusion the preformationist school—after the invention of the microscope, Malpighi could "still cherish the conjecture that...the [blood] vessels and the heart pre-exist and gradually come to view" as the embryo grows [Adelmann, 1966, p. 957]. The preexisting patterns may not always be apparent, these embryologists contended; but, visible or not, concordant and preexisting topological arrangements were always the mothers of succeeding embryological patterns.

In contrast, epigeneticists proposed that truly new configurations arise during embryogenesis: topology is often radically changed during embryogenic transformations. William Harvey ("who gave motion to the blood, even as he allotted to animals their origin") was an eminent seventeenth century embryologist as well as a physiologist and a physician; in his major embryological treatise, Harvey wrote—

> there is no part of the future foetus actually in the egg, but yet all the parts of it are in it potentially

and

> [all] the perfect animals, which have blood, are made by Epigenesis, or superaddition of parts.
>
> [quoted in Needham, p. 151]

And, a century later, Caspar Wolff declared: "He who defends the system of predelineation does not explain generation, but affirms that it does not exist!" [quoted in Gould, p. 34]. Epigeneticists claimed that the appearance of special ontogenetic order is not due simply to our lack of discernment, to our inability to perceive the preexisting patterns; rather, new organization arises where none preceded it. Many embryological patterns, says the epigeneticist, form without having been templeted by topologically similar precursors. Contemporary embryology suggests that the preformationist and the epigeneticist were both right.

II

Today, the three broad pattern analyses—definition, explanation, and action chronicling—answer the three elemental questions: What, How, and Whereto. Originally, however, explanatory analyses aimed to answer a different question: at one time, explanations answered "Why" not "How." This was Aristotelian explanation; Aristotle sought to explain the meaning of things, why things are. To Aristotle, explanation included the purpose and the design behind a phenomenon. With their undercurrents of purpose, intent, and meaning, "Why" explanations border on the psychological and even on the spiritual, and they can overstep the bounds between science and philosophy. In natural philosophy (as opposed to theology), any design underlying a phenomenon must ultimately be understood in human terms, and the commonly heard complaint that "Aristotle's explanations were teleological" comes from the unavoidable anthropocentricity of "Why" explanations.

Investigative science emerged during the Renaissance, and one of the innovations of this newborn science—a Galilean revolution—was the separation of descriptions of phenomena from the purpose, intent, or meaning of the phenomena. This revolution appeared most explicitly in the transition from "Why" explanations to "How" explanations:

> [Galileo Galilei (1564-1642)] had a few forerunners, a few friends sharing his views, but was strenuously opposed by the dominating philosophical school, the Aristotelians. These Aristotelians asked, "Why do the bodies fall?" and were satisfied with some shallow, almost purely verbal explanation. Galilieo asked, "How do the bodies fall?" and tried to find an answer from experiment, and a precise answer, expressible in

> numbers and mathematical concepts. This substitution of "How" for "Why," the search for an answer by experiment, and the search for a mathematical law condensing experimental facts are commonplace in modern science, but they were revolutionary innovations in Galileo's time.
>
> [Polya, pp. 194-195]

Embryologists are biology's explainers; and after the Renaissance, the central embryological question became a "How" question, specifically: "How do the special and stereotyped embryological patterns come about?" Preformationism and epigenesis posed two different a priori philosophical answers to this question. At the end of the nineteenth century, such a priori philosophical approaches were set aside, and in their place an investigative paradigm for answering this question was adopted. This procedure, the paradigm of alternative architectures, is a recipe that was developed in the nineteenth century school of *Entwicklungsmechanik* (developmental mechanics).

III

Understanding is built by doing: scientific meanings are operational. As A.S. Eddington wrote:

> The domestic definition of existence for scientific purposes follows the principle now adopted for all other definitions in science, namely, that a thing must be defined according to the way in which it is in practice recognised and not according to some ulterior significance that we imagine it to possess.
>
> [Eddington, p. 286]

Experimental paradigms are the operational role models of science, and paradigms develop hand in hand with technical advances. In the late 1800s, new questions, new perspectives, and new organizational explanations restructured Embryology as new techniques of embryological manipulation were introduced, and the prime mover behind these innovations was the German biologist Wilhelm Roux. Amphibian eggs, which are large enough for a wide range of surgical interventions, can develop from fertilization onwards in simple salt solutions; in 1894, Roux formalized the new school of experimental embryology by founding the journal *Archiv für Entwicklungsmechanik der Organismen,* and he simultaneously institutionalized amphibian eggs as the standard embryological system. Experimentally, Roux separated and killed individual embryonic cells to test the cellular independence of their inherent embryogenic plans; he also examined the effects of changing gravitational orientation, pressure, light, heat, electrical fields, and magnetic levels on the course of development, and he inferred that the point of entrance of the sperm into the egg determined the polarity of the embryo.

Technically, the new school of experimental embryology depended on the methods of tissue culture and of transplantation. These technologies fulfilled the requirements of Embryology in its role as the explanatory branch of Biology. Explanations describe how the parts of a thing come together to form a whole—thus, explanatory experiments comprise two technical tasks:

(a) discovering the intrinsic interactive potentials among the parts (the natural units) of a pattern,
(b) delineating the extrinsic information—the templet—that must be added to the intrinsic interactive potentials in order to produce the final pattern.

With their explicit emphases on the interactions among parts, the experimental paradigms of explanatory explanations are built of both isolation and combinatorial techniques. For embryos, the natural units are cells (Ross Harrison [p. 372] wrote: "the reference of developmental processes to the cell was the most important step ever taken in embryology"), and the explanatory paradigms of Embryology took form as soon as methods were invented to isolate and to recombine developing cells under conditions in which they would thrive. These methods are tissue culture for cell isolation and transplantation for cell recombination.

Tissue culture is the rearing of isolated cells and isolated groups of cells. For embryonic cells, tissue culture began with those animals that naturally develop in accessible loneliness. In the last half of the nineteenth century, the French embryologist Laurent Chabry showed that individual embryonic cells of the tunicate *Ascidiella aspersa,* a primitive marine chordate, will autonomously play out their natural fates; even when isolated from their companion cells, tunicate blastomeres develop into the appropriate differentiated parts of the whole organism. Soon thereafter, Roux reported similar results with isolated frog blastomeres.*

At the turn of the twentieth century, these new cell isolation techniques took on a sudden prominence because of a controversy that divided neuroembryologists. On the one side, Wilhelm His and Santiago Ramon y Cajal claimed that axons grow solely and completely out of neurons and that this growth is autonomous; in contrast, Theodor Schwann and Victor Hensen proposed that axons are formed largely by the concatenation of parts of nonneuronal cells. The evidence put forth by both sides came from fixed histological materials, and the actual dynamic events remained indirect inferences. To resolve the disagreement, Ross Harrison cultured embryonic neurons. Harrison grew cells from embryonic spinal cords, isolated in drops of culture medium (adult frog lymph), and under the microscope he observed directly the outgrowth of the axons. Axons, reported Harrison, extend completely, autonomously, and solely from neurons.

Tissue cultures are techniques for isolating; whereas transplantations are techniques for combining. In their most general terms, transplantations are physical rearrangements that produce new combinations of interactive elements. For example, in the late nineteenth century, Theodore Boveri centrifuged eggs of the nematode worm *Ascaris*; this centrifugal transplantation technique rearranged the

---

*In a striking contrast, Hans Driesch discovered that isolated sea urchin blastomeres do not develop into such restricted progeny: individual cells from young sea urchin embryos develop into complete adults, not merely partial adults, as would be expected if they had simply played out their normal in situ fates.

intracellular organization, bringing nuclei in contact with new areas of cytoplasm, and it produced correspondingly new developmental fates in the daughter cells. *Surgical* transplantation of groups of cells was introduced in the new school of experimental embryology by Gustav Born; in 1896, Born reported the chance discovery that cut amphibian embryos will reunite and heal and that even unnatural combinations—two embryonic trunk portions, for instance—can fuse, heal, and develop. Harrison and Hans Spemann immediately followed Born's observations and established surgical transplantation as part of a standard embryological paradigm. Victor Twitty, one of Harrison's students, described the situation this way:

> Spemann undertook the study of the first stages of development and, accordingly, the emergence of the body plan in its primitive outline, while Harrison dealt primarily with the analysis of subsequent differentiations as the special organs and tissues gradually took more definitive form. For the experimental manipulations of these early rudiments Harrison adapted or developed surgical techniques that were to be fruitfully exploited for decades.
>
> The parts of the embryo were soon parcelled out, like mining claims in a new gold field, among Harrison's students and followers, and eye, ear, limb, nervous system, and other parts, each became a surgical specialty yielding a rich bonanza of information. It is not difficult to understand why surgery was so peculiarly and profitably applicable to problems of growth and form. As development proceeds, the embryo becomes an increasingly complex assemblage of structures, all emerging at the "right" times and places relative to one another to form an integrated pattern or configuration that we call frog, bird, or mouse....If we disturb or rearrange the evolving system by excising selected parts or grafting them into new settings, we begin to expose some of the causal relationships and dependencies that were concealed. Since this is precisely what surgery, and surgery alone, permits the embryologist to do, what wonder that each spring the defenseless frog or salamander embryo has been assaulted on all fronts

> by investigators resolved to leave none of its parts inviolate or in their accustomed contexts.
>
> [Twitty, pp. 5-6]

The microsurgical manipulations introduced by Harrison and by Spemann allow both isolation and recombination. With surgical isolation techniques, the embryologist can methodically define minimal necessary sets of interactions; he does this by eliminating contacts that normally occur between cells or between parts of cells. By contrast, with surgical recombination techniques, the embryologist discovers potential but unrevealed interactions, and he does this by creating new combinations of cells or of parts of cells. With microsurgical techniques in hand, Harrison and Spemann forged a new experimental methodology for understanding developmental interactions, and the basic recipe for this methodology can be called the "paradigm of alternative architectures."

Alternative architectures sees each part of an organism as a particular pattern which develops from a predecessor embryogenic system. Embryogenic systems are collections of elements embodying the full potential to form the final patterns. In addition, some embryogenic systems also have the potential to form alternative final patterns. In these cases, the extrinsic information (the templet) that normally impinges on the system can be critically determinative: templets channel the diverse, intrinsic, and self-assembling potentials of embryogenic systems. The paradigm of alternative architectures explains a pattern by delineating an embryogenic system and its collaborative templet. For instance, among the patterns that we recognize in the tetrapod limb are the patterns formed by the digits of the hands. Each animal develops two distinct hand patterns: right and left limbs are mirror images of one another (Figure 7.1). We know the patterns of our hands as intimately as we know any biological pattern in the world, yet can we scientifically explain the configurations of our hands? How do these patterns come about? To explore this question, Ross Harrison applied the paradigm of alternative architectures.

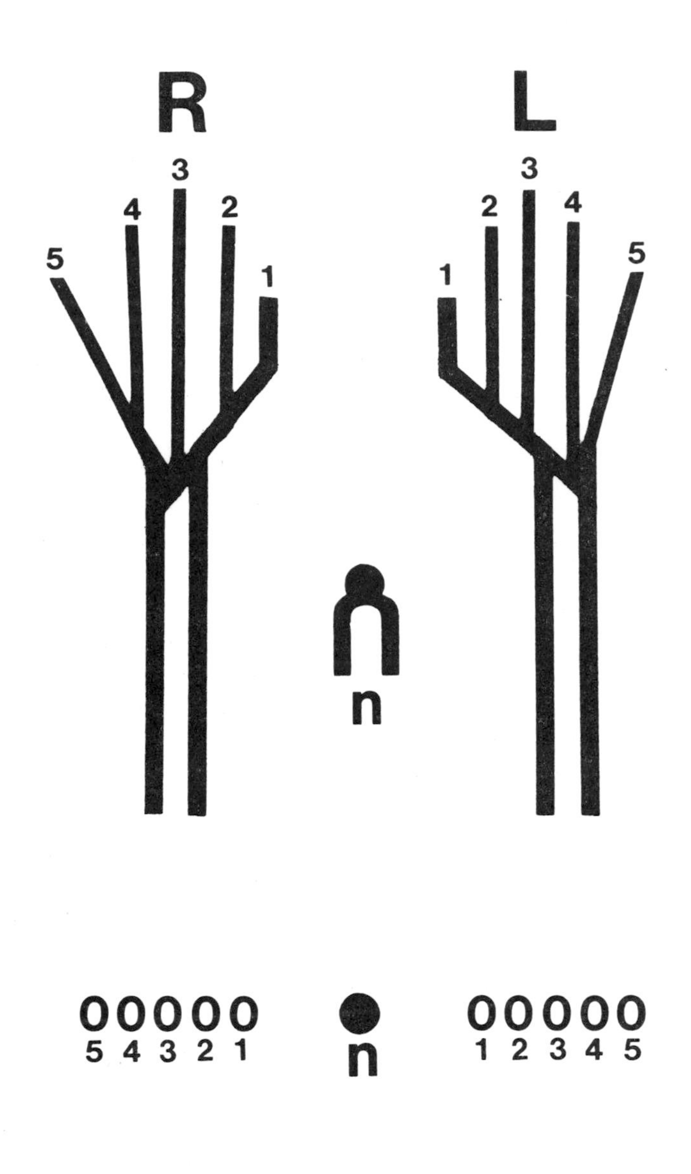

Figure 7.1

Schematic views of the tetrapod hands. Top: ventral views; bottom: frontal views. 1 = thumb, 2-5 = fingers, n = nose, marked for orientation to the overall body. (Harrison's classic experiments were on salamanders which have only *four* digits on each hand.)

To begin such an explanational analysis, one needs a well-defined object of study. Defining the pattern of study is an anatomical task,* and the full definition of any pattern includes a list of its content and its configuration. Harrison began his analysis by defining the overall pattern to be the hand, the unit elements to be the fingers, and the embryogenic system to be the small pile of cells—the limb bud—forming the limb primordium in the tail bud stage embryo of the amphibian.

To explain natural phenomena, we should work in natural units. Had Harrison partitioned the hand into natural units? To a human, fingers and toes feel like natural units, protruding, as they do, like independent sausages from the ends of our limbs. This feeling is primal: children as young as four years old include fingers as major individual units in their sketchy drawings of people. (This is about the time that children begin to add feet and ears to their drawings.) To Nature however, fingers are not entirely independent and determinate units (they are not simple, autonomous, all-or-none phenomena), because naturally-occurring developmental anomalies include digits that are fused, partially developed, shortened, lengthened, incompletely segmented, or barely identifiable tissue tags.

The somewhat indefinite nature of digits introduces fundamental problems for explanational analyses. Most importantly, digits themselves do not preexist as simple units within the limb's embryogenic system; thus, digits are not the most useful units for *alternative architectures* or other configurational explanations. A configurational explanation deals in pattern-assembly systems where content is essentially unchanged during the generation of configuration: the elements must preexist, and they must remain the same while a particular organization is being generated among them. Even the most preformationist aspects of contemporary Embryology recognize

---

*A prominent tool in the anatomist's laboratory kit is the decomposition-with-bookkeeping paradigm, which was elaborated by the molecular anatomist Frederick Sanger. (See chapter 6.)

only the inheritance of topology, not the inheritance of (most) complex structures like digits.

Preformed digits do not preexist in the limb bud embryogenic system, and ideally, then, the explanational analysis of the hand pattern should be built with more natural units. It is the cell that is the common and conserved unit of construction in metazoan organisms (this is the great insight of the Cell Theory of Mattias Schleiden and Theodor Schwann), and embryogenic systems comprise cells. Harrison knew this well, but he was not able to identify fully the appropriate cell patterns in the developing limb. In addition, he could not yet translate cell and tissue patterns into digit patterns: he did not understand the transformation of limb content. Thus, Harrison could not cleanly connect a configurational explanation of cellular paterns with the final digit pattern of the tetrapod hand. Today, we still need an explicit description of the transformation of limb bud content, from cells and tissues into digits—the deepest explanations are those that can be translated effectively into other parts, we need a translation into cellular units. Nonetheless, Harrison's analysis exemplifies the actual operation of *alternative architectures,* and it provides a fine window into practical configurational analysis.

Formally, a pattern analyst's explanation—a configurational explanation—of the form of a pattern *P* is $M \wedge T = P$. *M* is the matching matrix listing all of the inherent potentials (the "universal assembly laws") of the elements of *P, T* is the templet, and $\wedge$ denotes logical conjunction or Boolean Addition. Given a well defined pattern, the explainer's task is to discover the matching matrix and to infer the operant templet. The matching matrix completely defines an embryogenic pattern-assembly system, and the heart of the paradigm of alternative architectures is the disclosure of the matching matrix. What, asks the embryologist, are the complete potentials inherent in my embryongenic system, unfettered by particular situational constraints, freed of the cares of specific real world contexts? In this way, *alternative architectures* becomes a tool of science fiction—it probes worlds that *are not* but that *might be*. The scientific explainer deals partly in science fiction, because through its "universal assembly laws" an embryogenic system can embody all manner of fanciful configurations, some of which may never normally see the light of day.

A mother contains her child, and any pattern-assembly system must minimally contain the final pattern that it generates.The parts must at least contain sufficient interactive potentials to form the final whole—the elements must live in a world with "universal assembly laws" sufficiently encompassing to permit the generation of extant patterns. In the abstract formalism for a configurational explanation, the matching matrix ***M*** must contain ***P***. Therefore, this is the initial hypothesis presumed by the embryologist as explainer. Beyond these minimal conditions it is always possible that the elements of an embryogenic system can embody the potentials to form other patterns: many different final patterns may lie curled together, interwoven in their embryonic lineaments, within any given pattern-assembly system. To recognize any excess potentials, we must intrude upon our embryogenic system.

For the tetrapod limb, Harrison began by presuming the minimal matching matrix. To put the real experiments in a simple abstract form, let me schematize the hand pattern as composed of five elements—four digits (the thumb is digit 1) and one orientative marker *N* (the nose)—illustrated more generally in Figure 7.1. (Harrison worked with amphibians, which usually have only four fingers on their hands.) The amphibian hand pattern ***H*** can then be summarized in the following adjacency matrix:*

H =		N	1	2	3	4
	N	0	1	0	0	0
	1	1	0	1	0	0
	2	0	1	0	1	0
	3	0	0	1	0	1
	4	0	0	0	1	0

The appropriate minimal matching matrix is:**

---

*In this adjacency matrix, a *one* in a particular slot signifies that the two corresponding elements actually develop into adjacent structures. For example, the first row in H would be read as: only digit 1 normally develops adjacent to the nose end of the animal.

**In this matching matrix, a *one* in a particular slot signifies that the two corresponding elements have the *potential* to develop into adjacent structures. For example, the first row in M_0 would be read as: only digit 1 has the potential to develop next to the nose of the animal.

$$M_0 = \begin{array}{c|ccccc} & N & 1 & 2 & 3 & 4 \\ N & 0 & 1 & 0 & 0 & 0 \\ 1 & 1 & 0 & 1 & 0 & 0 \\ 2 & 0 & 1 & 0 & 1 & 0 \\ 3 & 0 & 0 & 1 & 0 & 1 \\ 4 & 0 & 0 & 0 & 1 & 0 \end{array}$$

Harrison began with this restrictive matching matrix and then asked whether the true inherent potentials are actually broader. Could the embryogenic system of the tetrapod limb also form other hand patterns?

All patterns develop within some real-world context, with its own particular opportunities, limitations, curfews, and sanctions. A purely self-assembling system is largely insensitive to these constraints and guides: self-assembling systems inherently generate stereotyped final patterns. In contrast, an extensively templeted system can generate variable patterns when the underlying templet varies. However, when the operant templet is normally invariant, then even templeted pattern-assembly systems will generate identical patterns time after time. Outwardly, the latter templeted systems resemble purely self-assembling systems, because in both situations the final patterns are recurrent and stereotyped.

How can the researcher recognize extensively templeted systems when the final pattern is normally stereotyped? Purely self-assembling systems have minimal matching matrices, but templeted pattern-assembly systems have less restrictive matching matrices. Therefore, to place a particular embryogenic system properly along the spectrum from purely self-assembling to completely templeted, one must intervene in the natural assembly process and vary the templet. A truly self-assembling system will be unresponsive to changes in the contextual constraints, whereas templeted systems can form different patterns in different environments.

The tetrapod limb is not perfectly self-assembling; some variant limb patterns occur naturally. Nonetheless, the limb pattern is highly stereotyped; less then 0.3 percent of children are born with any type of limb anomaly, and, with only this information, we cannot ascertain how close the embryogenic system is to an ideal, completely self-assembling system. To assess the true inherent potentials of the embryogenic limb system, the pattern biologist must intrude upon

Nature and actively vary the situational contexts in which the limb bud develops. Ross Harrision altered the situational constraints of this embryogenic system by transplanting limb buds to different locations and in differing alignments in salamander embryos. He began by collecting salamander eggs in the early springtime in streams of the New Haven countryside. After the embryos reached the tail bud stage, Harrison used watchmaker's forceps to pop them out of their jelly capsules and into finger bowls filled with a simple salt solution (optimally, about 6 millimolar NaC1 and 0.01 millimolar KC1). The forelimb bud is a 0.9 mm diameter bump on the side of the 4.3 mm long, stage 29 *Amblystoma punctatum* embryo. With the sharpened tips of small iris scissors, Harrison cut a limb bud from one embryo and fit it into a comparable-sized space excavated in a neighboring embryo; the graft was then held in place by the light pressure of tiny glass bandages, and the edges of the host and the graft healed together within an hour.

By changing the contextual circumstances of limb-bud development, Harrison created opportunities for new interactions, and he tested the inherent, but normally unrevealed, potentials for interaction between elements of the embryogenic limb system. A templet is contextual circumstances: it is the sum of the particular constraints and opportunities embodied by the specific real-world situation in which the creation of a pattern is played out. Thus, Harrison varied the templet to reveal the full matching matrix. The normal orientation of the limb bud aligns the thumb side of the hand toward the nose; Harrison reversed the limb bud, and he found that the thumb side of the hand can develop oriented toward the tail was well as oriented toward the nose. The actual matching matrix for the limb system is more permissive than is minimally necessary, so Harrision broadened his working model of the matching matrix to be:

$$M_1 = \begin{array}{c|ccccc} & N & 1 & 2 & 3 & 4 \\ \hline N & 0 & 1 & 0 & 0 & 1 \\ 1 & 1 & 0 & 1 & 0 & 0 \\ 2 & 0 & 1 & 0 & 1 & 0 \\ 3 & 0 & 0 & 1 & 0 & 1 \\ 4 & 1 & 0 & 0 & 1 & 0 \end{array}$$

Harrison then continued methodically to vary the opportunities for interaction among elements of the limb bud, and he followed the range of alternative architectures manifested by the system. For the embryological stage that he was testing, he found that the inherent associative potentials among the digit-forming elements were actually quite restrictive: the hand usually had the same internal pattern, and the digits formed in the same order. Harrison concluded that M_1 (above) is the full matching matrix and that it summarizes the "universal assembly laws" of the digit pattern of the amphibian hand.

Pattern-assembly systems are always tempered by real-world constraints, curbs, and guides that together form a templet channelling the inherent self-assembly potentials. After exposing the "universal assembly laws," the embryologist infers the class of operant templets, and with all the components summarized as matrices, the inference is a straightforward exercise that follows three rules. For this step, the researcher can settle back in his armchair and temporarily become a theoretician. In the abstract, one simply overlays the adjacency matrix onto the full matching matrix, and:

(1) In those slots where a 1 in the matching matrix overlays a 0 in the adjacency matrix, an interactive potential of the pattern-assembly system has been curtailed. The corresponding slot in the templet matrix summarizes this constraint with a 0.

(2) In those slots where a 1 in the matching matrix overlays a 1 in the adjacency matrix, an interactive potential of the embryogenic system has been realized in the final pattern. The corresponding slot in the templet matrix summarizes this opportunity with a 1.

(3) In those slots where a 0 in the matching matrix overlays a 0 in the adjacency matrix, no interactive potential exists in the embryogenic system and the templet has no influence on the final pattern. Here, it is of no consequence whether the templet offers either a constraint or an opportunity. The corresponding slot in the templet may contain either a 0 or a 1. (We can summarize this indefinite state by an "x".)

In this way, the pattern biologist builds his class of candidates for the actual operant templet. The appropriate templets embody a set of necessary constraints and opportunities (determined by Rules 1 and 2), and often they will also have unnecessary constraints or opportunities (as identified by rule 3). Nature need not follow human extremum principles, such as the principle of parsimony (Occam's Razor), and the true operant templet cannot always be completely inferred by armchair theorizing. To fill in the actual details for those matrix slots identified as x's by Rule 3, the researcher must again metamorphose from the theoretician back to the natural historian or the experimentalist.

With this logic, Harrison used his discovery of the "universal assembly laws" of the tetrapod hand to infer the nature of the still hidden templet. For our schematic hand pattern, the appropriate class of templets takes the form:

$$T = \begin{array}{c|ccccc} & N & 1 & 2 & 3 & 4 \\ \hline N & 0 & 1 & x & x & 0 \\ 1 & 1 & 0 & 1 & x & x \\ 2 & x & 1 & 0 & 1 & x \\ 3 & x & x & 1 & 0 & 1 \\ 4 & 0 & x & x & 1 & 0 \end{array}$$

In narrative form, this model templet proposes that the situational context within which the limb bud system generates a hand contains:

(a) opportunities (1's) for the digits to develop adjacent to one another and also an opportunity for the thumb to develop on the nose side of the limb,

(b) constraints (0's) that disallow the last digit from developing on the nose side of the limb, and

(c) unrestricted interactions (x's) which have no influence on the associations of the digits in the final pattern.

Only further experimentation can properly fill the unrestricted slots (x's) in this model templet; in a world that Nature has wrapped with incondensable complexity, in a universe chocolate-thick with variety and superfluity, we cannot make confident a priori guesses. The full explanation of a pumpkin seed is found in the lopsided pumpkin overgrown with vines and splashed with mud, and not in our comfortable but contracted and ultimately oversimplified living rooms.

IV

In carrying out the paradigm of alternative architectures, Harrison assayed the fates of limb buds developing in a wide variety of contexts. At early embryological stages, the results of transplantation were the same as at late embryological stages—the underlying assembly laws always permit the development of two opposite hand patterns;

N-1-2-3-4 and 1-2-3-4-N

For this particular aspect of hand patterns, the matching matrix appears to remain invariant during the development of the amphibian. In contrast, the matching matrix changes for another aspect of the hand pattern. At early embryological stages, the palm side of the hand develops only adjacent to the belly side of the animal, regardless of the initial orientation of the limb bud. At late embryological stages, however, the palm side of the hand can also develop adjacent to the back side of the animal: late-stage surgical interventions can make the hand develop upside-down. For "palmness," the inherent assembly laws of the embryogenic limb system become more permissive as the animal develops.

The matching matrix fully describes an embryogenic system—to know the embryonic elements and their inherent associative potentials is to know a pattern-assembly system. The "palmness" pattern illustrates that during development, the matching matrices can also evolve, and this means that embryogenic systems can change as the adult organism unfolds. Embryogenic systems are transformed; at any one time, new and characteristic pattern-assembly systems become the active players, and earlier embryogenic systems remain only as intangible memories. Earlier stages are like the summer sunset bird that Robert Frost remembered again in winter, no longer substantial but nonetheless touching us in the present.

> In summer when I passed the place
> I had to stop and lift my face;
> A bird with an angelic gift
> Was singing in it sweet and swift.
>
> No bird was singing in it now.
> A single leaf was on a bough
> And that was all there was to see
> In going twice around the tree.

Throughout development, embryogenic systems transform like clouds in the sky—first we find an irregular oblong, the oblong rolls into a limb bud, and then it stretches into a hand. These pattern systems continually sweep through the ever-changing organism, and in any one creature they quickly become intangible. Unlike the clouds, however, embryogenic systems recur, and we can watch them again and again in our children, in our grandchildren, and in all of their never-ending descendants.

> So [Christopher Robin and Winnie the Pooh] went off together. But wherever they go, and whatever happens to them on the way, in that enchanted place on the top of the Forest, a little boy and his Bear will always be playing.
>
> [Milne, pp. 179-180]

V

A recipe is a naked description, generic instructions for building things, routine and well-defined operational maneuvers that anyone can follow. Recipes can be thoroughly examined and tested by many different researchers, and each examiner will inevitably apply his own unique and critical eye. The contemporary recipe form is the algorithm, and as a recipe, *alternative architectures* is:

THE ALTERNATIVE ARCHITECTURES ALGORITHM

MAIN PROGRAM

Define the pattern

(apply the *Decomposition-with-Bookkeeping* algorithm)

Determine the associative potentials between elements: build a matching matrix

Begin with the minimal matching matrix as a working hypothesis

Test the pattern-assembly system by varying the situational constraints

1. **Methodically create opportunities for new interactions**
2. **Broaden the matching matrix when new interactive potentials are discovered**
3. **Repeat steps 1 and 2 until all possible interactions have been tested**

Propose a templet using the three rules of templet inference

Build a model

Draw the matching matrix, test the model, and remodel

Draw the templet, test the model, and remodel

Chapter Eight

HAUNTS-AND-CONSEQUENCES

The third broad analytical operation of the pattern biologist is chronicling. *This is a characteristic activity of Physiology; it is prospective, causal analysis, and it uses the haunts-and-consequences paradigm.*

I

Tangled gears springing teeth
Wheels pulleys underneath
Ticking of the ratchet clockwork
Clanking chimes in tiny doll kirk
Rolling bowing homunculi—
And all were wound by little I.

Efficient mechanical clocks were invented in the thirteenth century, when:

> No less important than its value as a timekeeper was the inspiration [that mechanical clocks] gave to ingenious minds to inquire whether the same principles could not be applied to different fields... The excitement generated by a machine which ran of itself with one small boy to tend it can hardly be appreciated today. The automatic machine seemed to have more than a little of the miraculous about it.
>
> [Keller, p. 13]

Clocks are automatic pattern-generating machines, and mechanical clocks are complex but comprehensible contraptions with tangible parts. At the same time, they are autonomous: clocks can unfold myriad and kaleidoscopic patterns in a spontaneous sequence. Physiologists see animals as clockwork machines; and this physiological vision arose in the Renaissance, because mechanical clocks were not common Europe until the fourteenth century. The Greeks and the Romans had used the clepsydra, a water clock with few mechanical parts. Machines in general, and especially those machines like mechanical clocks with many and interactive parts, played no significant role in Greek or in Roman life; for this reason, machine metaphors for organismal function were rare in pre-Renaissance science.

Physiology is the science of organismal function. Initially, Physiology was chemical; in Greek and Roman science, body parts and body functions were considered to be combinations of the four elements *(fire, air, water,* and *earth),* the four qualities *(hot, dry, wet,* and *cold),* and the four humors *(blood, yellow bile, phlegm,* and *black bile).* (The Galenic system of medicine explained diseases as inappropriate proportions of these twelve elemental components.) Throughout the Middle Ages, Physiology remained chemical and separate from mechanics. The machine tradition of Western science had grown out of the framework set by the Greek geometer and mechanical engineer Hero of Alexandria (the second century AD). Hero had formulated an elemental analysis of machines—specifically, he regimented the mechanical elements to comprise five simple components (wheel, lever, pulley, wedge, and screw), and his classification system for all machinery held sway until the beginning of the nineteenth century.

Those few machine analogies that did arise for organismal functions followed directly in the spirit of Hero of Alexander. In *On the Usefulness of the Parts of the Body,* Galen described the lower end of the humerus as a pulley; he suggested that the aorta also acts as a pulley or "turning-post" for the recurrent laryngeal nerve when the neck is turned. Galen equated the movement of our joints to those of marionettes, and he compared the larynx to a flute: the voice, he wrote, is produced by the movement of air through a narrowed aperture, just as music is produced in an instrument. In addition, Galen introduced one machine metaphor (a mechanical model for the circulatory system) that lasted more than fifteen centuries:

> The heart itself...snatches and, as it were, drinks up the inflowing material, receiving it rapidly in the hollows of its chambers. For [think if you will] of a smith's bellows drawing in the air when they expand, [and you will find] that this action is above all characteristic of the heart.
>
> [May, p. 316]

In the Renaissance, this model was reiterated and expanded by Leonardo da Vinci, who compared the heart to a bellows furnace, and by Rene Descartes, who compared the heart to the air bellows of a church organ.

In spite of Galen's metaphors, mechanical understandings of animal functioning remained a minor theme until the Renaissance. Then, machines invaded our day-to-day lives. It was Leonardo da Vinci who embodied the new physiologist's view—a machine view—of organisms: "Just as Leonardo the artist focused his attention on the emotional movements of man, so Leonardo the engineer focused on man the machine." [Keele, p. 163] In Milan during the first two decades of the sixteenth century, da Vinci "returned to his researches on the structure and function of the human body...making a full-scale study of the whole body as a mechanical instrument." [Keele, p. 37]

Throughout his many treatises, da Vinci repeatedly used the word "machine" as a synonym for body. (For example, from his annotation of a set of anatomical sketches: "O observer of this machine of ours, let it not distress you that you give knowledge of it through another's death, but rejoice that our Author has established the intellect in such an excellent instrument." [Keele, p. 288]). Da Vinci listed the fundamental "18 operations of man" as: "rest, movement, running, standing, supporting, sitting, leaning, kneeling, lying down, suspended, carrying or being carried, pushing, pulling, beating, being beaten, pressing down, or lifting up"; he then founded bio-mechanical physiology by analyzing each of these human operations into the compounded actions of levers and balances. For sensory physiology, da Vinci proposed perception by reception—we perceive by receiving emissions from things, not by sending out some "receptive spirit" like a radar; we see a flame by capturing light sent from a candle, and we hear a clang by capturing sounds sent from a bell. He also introduced a machine model of the eye, the anatomy and function of which, he suggested, are directly analogous to the structure and operation of the camera obscura.*

In the sixteenth century, da Vinci imbued biology with a thoroughly machine perspective; and in the seventeenth century, Descartes emphasized the specifically *clockwork* aspects of this perspective. Descartes considered our body "to be but a statue, an earthen machine [like] clocks, artificial fountains, mills, and similar machines," [Descartes, pp. 2-4] and he expanded his metaphor with his famous comparison:

> Moreover, breathing and other such actions...are like the movements of a clock or mill which the ordinary flow of water can render continuous. External objects which merely by their presence act on the organs of sense and by this means force them to move in several different ways, depending on how the parts of the brain are arranged, are like strangers who, entering some of the grottoes of these fountains, unwittingly cause the movements that then occur, since they cannot enter without stepping on certain tiles so arranged that, for example...they will make a marine monster come out and spew water into their faces, or other such things according to the whims of the engineers who made them.
>
> [Descartes, p. 22]

Descartes explored all of biology in terms of clockwork organisms, built mechanically from tiny animated units; to Descartes, Physiology was "the mechanics of very small things [in motion, beginning] just below the level of vision and [reaching] downward to the level of elementary particles of matter." [Descartes, p. xxix] Even in his atomism, however, Descartes codified the physiologist's insight: animal physiology stems directly from an animate machine, composed

---

*The camera obscura is a device with a small hole or lens that passively projects images onto a darkened surface.

of tangible parts. Autonomous, manifold, and kaleidoscopically wondrous as organismal functioning is, it is nonetheless comprehensible, because animal activities are patterns generated by clockwork organisms.*

Biology is operational: it is a manual science populated by artisans, artificers, and activists. Theoreticians must stand aside, because the world of organisms is sufficiently varied, idiosyncratic, and peculiar that direct involvement with living things is the only way to build up complete understandings. Biologists immerse themselves in nature—they reach out into the thick tangle of uniquely specialized organic architectures to learn the full extent of complex living patterns and to find the unexpected corners and the abrupt edges of cytoplasmic extensions and their surrogates.

Operationally, Physiology uses paradigms devised for Anatomy and for Embryology, but it also has its own special experimental role models. Physiology focuses on organismal functioning, on living processes, on biological action. The major biological actons are: circulation, digestion, metabolism, movement, perception, respiration, and water regulation, and the uniquely physiological paradigms are paradigms designed to reveal these biological actions alive in their natural habitats. The biological actions are animate patterns transpiring through the concordant integration of many elements; they are products of a complex clockwork of living machinery, and they emerge synergistically. Thus, the uniquely physiological paradigms cannot simply take the systems apart and list the separate components, as do the anatomical paradigms. Moreover, Physiology is concerned with those particular processes that actually happen in nature. Thus, the special paradigms of Physiology cannot stray into science fiction, as does the paradigm of alternative architectures of Embryology. Instead, physiologists have the challenge of seeing the full biological actions and actors alive and interacting in their own homes, playing

---

*In *Man A Machine*, Joseph Needham follows the history of the mechanistic versus the vitalistic traditions of Physiology beyond Descartes and into the early twentieth century.

together at their own hearths, and breathing vitally throughout their own native haunts.

Physiologists study living systems that are actively playing out their biological actions, and the most direct physiological approach is external observation. For instance, William Harvey inferred that blood circulates in a closed loop within the body, and a key step in Harvey's logic was his direct observation that the valves in veins keep blood moving unidirectionally, toward the heart. Venous valves had been discovered earlier, in the 1540s, by Giovanni Canano and by Jacobus Sylvius, and they were described to Harvey during his anatomical studies with Fabricius. However, various incorrect functions had always been ascribed to the venous valves. It was only by direct observation that Harvey convinced himself of the valves' true biological action; Harvey put a tourniquet around patients' arms, and he found that blood accumulated in veins on the side *away* from the heart and on the proximal side of the valves (''nodes''). Moreover,

> if by milking the vein downwards with the thumb or a finger you try to draw blood away from the node or valve, you will see that [no blood] can follow your lead because of the complete obstacle provided by the valve...If you keep the blood thus withdrawn and the vein thus emptied, and with your other hand exert a pressure downwards towards the distended upper part of the valves, you will see the blood completely resistant to being forcibly driven beyond the valve..
>
> [quoted in Fulton and Wilson, pp. 51-52]

Useful as they are, such direct external observations are nonetheless quite limited, because most physiological processes are hidden within organisms—behind shells, under skins, within vessels, beyond membranes, or beneath the resolution of the unaided human eye. Therefore, physiological paradigms usually involve interventions that dissolve these barriers while still preserving the complete biological action. Dissection, for instance, can maintain much of the system intact while opening it up for us to look inside. The introduction of dissection as an experimental paradigm is usually attributed to the Greek physician Alcmaeon of Croton, (a student of Pythagoras), in about 500 BC. "Alcmaeon has been called the earliest Greek physiologist since he pioneered animal experimentation, carrying out vivisection experiments in order to solve biological and medical questions." [Rothschuh, p. 3] Later, the Roman physician Galen based his physiology on the dissection of live animals, having performed such in situ experiments as cutting the recurrent laryngeal nerve in a squealing pig to demonstrate that the brain and not the heart or the lungs controls the voice. In the classical eras, dissection of human cadavers went through periods of cultural acceptability, and human physiology was then indirectly built from comparisons of human dissections and animal vivisections.

Direct human experimentation has always carried the weight of a special emotional cast. Lazaro Spallanzani was the eighteenth century Italian naturalist who first demonstrated that adult newts can repeatedly regenerate amputed limbs; also, by carefully boiling organic fluids in closed flasks, he temporarily put to rest scientific claims of the spontaneous generation of life. To explore human digestion, Spallanzani swallowed cloth bags of various foods and studied the remnants after they had passed through his system. With this program, Spallanzani showed that human digestion was similar to digestion in other carnivores; among his specific findings were that complete digestion does not take place in the human stomach and that human stomach juices work differently on different types and different preparations of foodstuffs.

More recently, another direct human experimental paradigm has been the physiological exploration of the organization of the human brain. Using tiny electrical stimuli in awake patients during neurosurgical operations, physiologists have mapped out the functional components of the cerebral cortex. The Canadian surgeon Wilder Penfield, who pioneered this work, has written eloquently of "how I listened to the humming of the mind's machinery, and where, [for example, our] words come from." [Penfield and Roberts, p. x]. In Penfield's research:

> local anaesthesia was used during the operations...This does away with the pain of the procedure and yet leaves the brain normally active after a segment of the skull has been cut and temporarily turned back and the surface of the brain

> thus exposed...Since the patients were talking and fully conscious during the procedure, it was possible to discover what parts of the cortex were devoted to [particular] functions.
>
> Take as an example the words of M.Ma. during stimulation of her right temporal lobe. The brain had been exposed [and] when an electrode, insulated except at the tip, was introduced...one centimeter into the cortex of the superior surface of the temporal lobe and a gentle current was switched on, she exclaimed:
>
> "Oh, a familiar memory—in an office somewhere. I could see the desks. I was there and someone was calling me—a man leaning on a desk with a pencil in his hand."
>
> All the detail of these things to which she had paid attention in some previous period of time were still there.
>
> [Penfield and Roberts, pp. 5 and 45-46]

Penfield's work in nervous systems is in the tradition of the physiology of *whole* systems. A complementary physiological approach sorts out the multitude of simultaneous effects and mechanisms by working with *partial* systems. The nervous system is the most interdependent of all metazoan tissues; to understand its operation, physiologists have studied various partitionings, searching for subsystems that retain particular biological actions. For example, early in the eighteenth century, Alexander Stuart separated the brain from the spinal cord in frogs and showed that many reflexes and even coordinated jumping movements and "defensive actions" are intrinsic to the spinal cord itself. Then, in 1822, the French physiologist Francois Magendi demonstrated that sensory aspects of coordinated behavioral actions are anatomically separate from motor aspects; specifically, the dorsal roots of the spinal cord selectively transmit sensory information, while the ventral roots selectively transmit motor information.

The final extension of analysis by partitioning involves studies of reconstituted systems. Here, known ingredients are mixed together in measured quantities in order to synthesize a system that recreates specific biological actions and that reiterates particular biological patterns. This is a more controlled but a physiologically riskier approach. In reconstituted systems, key elements are introduced by the experimenter from the outside; thus, even when the reconstituted system mimics natural actions, one must verify that the processes in the reconstituted system are identical with those particular processes that actually bring about the natural actions in situ. For instance, to explore the role of electricity in coordinated animal movements, Aloisius Luigi Galvani showed that electrical sparks could produce reflexlike movements in the decapitated frog preparation.* From his reconstitution experiments, Galvani showed that electricity was the natural currency of animal movements, but it remained to be demonstrated that this potential actor was in fact the actual actor in the natural setting. The physiologist's final question is not "What are the possibilities"—his question is "What are the actualities?" Therefore, to generate a complete physiological understanding, reconstitution experiments—analyses of the effects (the consequences) in controlled systems—must be balanced by natural history experiments—analyses of the indigenous occurrences (the haunts) in natural systems. The operational role models unique to Physiology combine *both* of these analyses, and for this reason such studies can be called "haunts-and-consequences paradigms."

III

> Where am I going? I don't quite know
> Down to the stream where the king-cups
> grow—

---

*Galvani's experiments were precipitated initially by a strange meteorological effect:

> I had upon occasion remarked that prepared frogs, which were fastened by brass hooks in the spinal cord to an iron railing which surrounded a certain hanging garden of my home, fell into...contractions...when lightning flashed.
>
> [quoted in Fulton and Wilson, p. 226]

> Up on the hill where the pine trees blow—
> Anywhere, anywhere. *I* don't know.
>
> [A.A. Milne]

There are three understandings that can be created for any biological pattern: a pattern can be defined, it can be explained, and its actions can be chronicled. *Definition* is a pattern's present, *explanation* is its past, and the *action chronicle* is its future. These three time frames are distinct visions of a pattern. In the present, a pattern lives independently, with no regard to its past or its future. From the past, the pattern emerges; and in the natural universe, unconstrained by extremum principles of construction—extremum principles like parsimony, simplicity, efficiency, versatility, excellence, or elegance—various ancestral pedigrees can converge to the same end. The means always explain the end, but the end need not explain the means. Into the future, the pattern rolls on blindly. The future is a cone broadening out into times aeonian; many available roads diverge from the present. "Whereto" and "Whither" are the pattern questions of travel along these roads, and for Physiology their answers constitute a travelogue and a chronicle, not a prediction.

IV

Our scientific world takes its meaning from physical operations that we carry out, and

> two aspects of..."meaning" are involved. [First, on a personal level] I am never sure of a meaning until I have analyzed what I do, so that for me meaning is to be found in [an explicit] recognition of the activities involved...[Second, on a scientific level] the operations which give meaning to our physical concepts should properly be physical operations, actually carried out.
>
> [Bridgman, pp. 8-9]

Operationally, Physiology chronicles the natural travels of particular biological entities. The general methodology for following biological objects is vital staining, and improvements in vital staining techniques have stimulated advances in Physiology. "To stain" is selectively to make a specific entity more apparent; "to vitally stain" is selectively to make a specific biological entity more apparent in situ without disrupting the natural physiological actions of the system. With this definition, microscopes might be considered vital stains, because they are a technology for selectively making apparent those biological entities too tiny to be resolved by the unaided eye. Magnifying devices have been used since ancient times—hollow spheres of glass were common magnifiers in Rome—even so, the introduction of the microscope as a scientific tool, like the introduction of machines and much other technology, was a Renaissance event. "Galileo had already invented his telescope in 1610, in order to observe insects and other structures invisible to the naked eye," [Rothschuh, p. 186] and Descartes described a microscope in his *Dioptrique* of 1637. Concurrently, the Dutch naturalist Antony von Leeuwenhoek succeeded in grinding and polishing lenses of exceedingly short focal lengths; some of his lenses magnified objects 270 times, making the simple microscope into a powerful biological tool.

Von Leeuwenhoek was a true physiologist, perhaps the first physiological microscopist; he reported on observations of tiny blood vessels in tails of living tadpoles, in eels, in frogs' feet, and in bat wings, and he described living spermatozoa and also a variety of living bacteria and protozoa. His in situ studies, and the comparable work of the Italian anatomist Marcello Malpighi, directly demonstrated the self-contained blood circulatory system postulated by William Harvey: von Leeuwenhoek and Malpighi discovered the blood capillaries. As von Leeuwenhoek wrote:

> I observed the young frogs [and] discovered in them a very large number of small blood-vessels which, continually running in curves, formed the vessels called arteries and veins, from which it was perfectly clear to me that the arteries and veins are one and the same continuous blood-vessels.
>
> [quoted in Fulton and Wilson, pp. 73-74]

At the same time, Robert Hooke, Curator of Experiments for the Royal Society of London, wrote his famous *Micrographia, or Some Physiological Descriptions of Minute Bodies made by Magnifying Glasses* (1665) from observations with compound microscopes that magnified objects thirty times. Accompanied by beautiful and detailed drawings, Hooke's descriptions of living material (for instance, his discovery that common mold is actually a lawn of tiny blue and white balloons on slender stalks) fascinated scientists and philosophers of his day. Hooke properly titled his book "Physiological," for he included vital staining descriptions—that is, descriptions of living organisms in their natural states selectively "stained" with magnification.* Hooke's full *Micrographia* is a compendium of microscopical observations of cloth, sand, snowflakes, plants, seeds, insects, feathers, urine, metal, and crystals, a discussion of optics

---

*For instance, Hooke wrote:

> Reading one day in *Septemb.* I chanced to observe a very small creature creep over the Book I was reading, very slowly; having a *Microscope* by me, I observ'd it to be a creature of very unusual form...It was about the bigness of a Mite, or somewhat longer, it had ten legs, eight of which were topt with very sharp claws, and were those upon which he walk'd, seeming shap'd much like those of a Crab, which in many other things also this little creature resembled; for the two other claws, which were the foremost of all the ten,...seem'd to grow out of his head [and] were exactly form'd in the manner of Crabs or Lobsters claws, for they were shap'd and jointed much like those...and the ends of them were furnish'd with a pair of claws or pincers, which this little animal did open and shut with pleasure...In all likelihood, Nature had crowded together into this very minute Insect, as many, and as excellent contrivances, as into the body of a very large Crab, which exceeds it in bulk, perhaps, some Millions of times...It being a general rule in Nature's proceedings...that there is no less curiosity in these parts which our single eye cannot reach, then in those which are more obvious.
>
> [Hooke, pp. 207-208]

and astronomy, a set of respiratory and atmospheric experiments, and also the first description of cells.

To the biologist, vital staining usually connotes vital dyes. Vital dyeing—highlighting selective living tissues—predated histological staining in general. Originally, the most common vital dye was madder root (the Eurasian herb *Rubia tinctorum*), which contains two members of the oxyquinone group of dyes: sodium alizarin sulphonate (alizarin red S) and purpurin-3-carboxylic acid; these two dyes have a predilection for newly deposited bone calcium, and animals that eat madder plants develop pink bones and teeth. Scientific references to madder dyeing date back to the sixteenth century, and in the mid-eighteenth century, physiologists used madder as a vital dye for studying natural bone growth in situ. For cell studies, techniques of vital dyeing developed along with the art of nonvital histological staining, late in the nineteenth century. At that time, several general dyes, "such as methylene blue and neutral red, were found to be of such low toxicity that they could be applied to cells which are still alive," [Hughes, p. 121] and some of the vital dyes, such as carmine and colloidal carbon particles for phagocytic vesicles and Janus green for mitochondria, were also found to be specific for particular organelles. Today, the most powerful vital dyes are fluorescent- or radioactive-tagged molecules; these dyes mix carefully engineered specific and physiologically-innocuous core molecules with the potential for high amplification of the fluorescent or the radioactive signal. To search out the haunts and to follow the consequences of a biological entity, the physiologist needs such vital dyes among his tools.

V

In the concordantly harmonious universe of the Pythagorean tradition, man is a microcosm—he is a literal miniature of the universe, which is the macrocosm—and all phenomena within man are majestically parallelled outside of him. If man is a microcosm, then his brain must be an ultramicrocosm; for, with its tentacular nerve net permeating and mirroring the body and with its multiple topographic, homuncular maps, the brain is a collection of miniatures of the body. The brain is an ultramicrocosm, and it is as thickly complex as the macrocosm. Thus, a full scientific understanding of the brain would be complex, thick, and tangled; it would be difficult to comprehend and impossible to summarize in simple apothegms or formulas. Nonetheless, humans deal best in simple understandings, and scientists strive against the inherent complexities of the natural world to make their explanatory statements simple. Physiologists found it possible to simplify their understandings of the brain when they discovered that many of its fundamental actions are preserved in partial systems. In 1739, for example, the English physician Alexander Stuart demonstrated that a variety of coordinated movements could be fully enacted by the spinal cord alone in decapitated frogs.

A century later, Marshall Hall—the London neurologist who emphasized the primacy of direct observation ("The sources of our knowledge in physiology as well as in natural science are *observation* and *experiment*")—was the first to demonstrate the autonomy of the local reflex arc, and he emphasized the importance of the local and the involuntary components of many coordinated actions of skeletal muscles.* As a mechanistic basis for these local behaviors, Galvani had shown earlier that electrical stimulation can produce natural reflex action, eliciting gastic secretion. Next, Carlo Matteuci in Italy and Emil Du Bois-Reymond in Germany showed that normal muscular contractions are accompanied by regular electrical depolarizations. All of these observations supported the contention

---

*Hall read his experiments before the Royal Society of London in 1833, concluding:

> It is distinctly proved, by this series of observations, that the reflex function exists [locally] in the medulla independently of the brain; in the medulla oblongata [i.e., hindbrain] independently of the medulla spinalis [i.e., spinal cord]; and [locally] in the spinal [cord] of the anterior extremities, of the posterior extremities, and of the tail, independently of that of each other of these parts, respectively.
>
> [Hall, p. 665]

This engendered such derision that the *Philosophical Transactions of the Royal Society* refused to publish further papers of his.

that electrical waves are synonymous with excitation throughout the nervous system; and at the end of the nineteenth century, electricity was considered to be the sole and common currency of the nervous system. The same chemical reactions that propagated the electrical stimuli down an axon, it was thought, also transmitted the stimuli through the synapses both between neurons and other neurons and between neurons and effector organs, such as glands and muscles. This concept was enunciated by the German physiologist Willy Kuhne,* who directly demonstrated that the natural electrical waves in muscle are fully sufficient to produce contractions in other muscles and are also sufficient to stimulate nerves.

Nonetheless, a collection of observations remained at odds with entirely electrotonic synapses. In 1845, the German physiologists Ernst and Eduard Weber reported that electrical stimulation of the severed vagus nerve will decrease or stop heartbeats: this showed that reflex neural pathways could be inhibitory as well as excitatory. The Webers also found that repeated electrical stimulations become decreasingly effective:

> If excitation of the vagus nerves is continued long enough for the force producing their excitation to be exhausted, the heart begins to beat again.
>
> [quoted in Fulton and Wilson, p. 296]

Electrical stimuli were not continuously effective—in itself, electricity was not the full and sufficient cause of nerve action. Could there be another intermediary actor? Claude Bernard had shown that curare, the famous South American arrow poison, blocked neuromuscular

---

*Kuhne (1837-1900) was a student of the French physiologist Claude Bernard. He was the first scientist to describe the motor endplate in muscles, and he discovered and named the ubiquitous contractile protein *myosin;* he also introduced the term "enzyme," and he named the specific pancreatic enzyme *trypsin.* In a dramatic demonstration of the chemical substrate of vision, Kuhne used an alum solution to fix and to preserve the image (an "Optogramm") of a window on the retina of a rabbit.

transmission without affecting the muscles' sensitivity to direct electrical stimulation: Bernard found that synaptic transmission could be separated from the electrical propagation of impulses. These findings complicated the arguments in support of simple electrotonic synapses; but direct proof of another possibility—specifically, chemically-mediated synaptic transmission—was still needed.

Henry Dale was trained in the English school of endocrinological physiology, and he viewed the natural chemical acetylcholine (Figure 8.1) as a neural hormone. In 1914, Dale demonstrated that the administration of acetylcholine mimicked the stimulation of parasympathetic nerves, and he distinguished two classes of acetylcholine effects: muscarinic and nicotinic. These two effects were named after two compounds—the alkaloids muscarine and nicotine—that could evoke selectively the two different actions. Both acetylcholine and muscarine act on smooth muscles, heart muscles, and gland cells, and these effects are counteracted by atropine; in contrast, both acetylcholine and low doses of nicotine act on autonomic ganglia, adrenal medulla cells, and motor endplates, and these effects are counteracted by curare and by high doses of nicotine.

Then, beginning in 1922, the physiologist Otto Loewi published a set of papers that gave:

> the direct experimental evidence of the...chemical transmission of the excitatory (or inhibitory) process, from the endings of vagus and sympathetic nerve fibres to the heart muscle of the frog. The experiments had a classical directness and simplicity. The washed heart...held a small quantity of Ringer's solution. When the appropriate nerve was effectively stimulated, this solution acquired the property, when removed and transferred to a second, control heart, of reproducing in this the effect of the nerve stimulation on the first, whether inhibition by vagus or acceleration by sympathetic impulses.
>
> [Dale, 1953, pp. 614-615]

The inhibitory property was due to a substance ("vagusstoff") that was later indentified as acetylcholine, and the excitatory humor ("accelerans-stoff") was norepinephrine. Similar experiments by

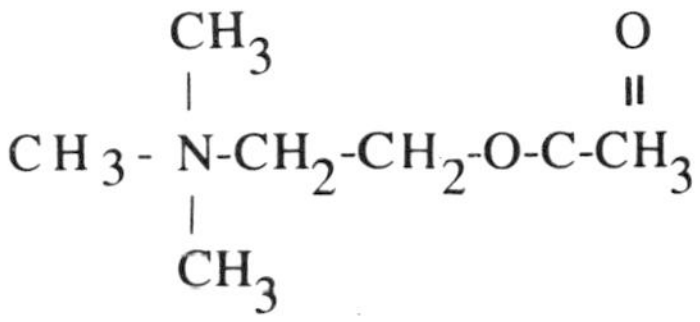

$$CH_3-\overset{\overset{CH_3}{|}}{\underset{\underset{CH_3}{|}}{N}}-CH_2-CH_2-O-\overset{\overset{O}{\|}}{C}-CH_3$$

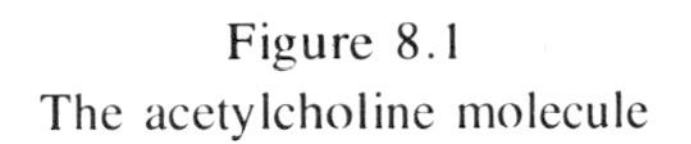

Figure 8.1
The acetylcholine molecule

Walter Cannon and his colleagues reproduced Loewi's main result in mammals and then correlated these chemically-mediated effects of sympathetic nerve stimulation with the emotional effects of stress. Thus, in the first quarter of this century the stage was set: chemically-mediated synaptic transmission was likely to be a widespread neural mechanism, and acetylcholine was a candidate for principal actor. It then remained to document this scenario in natural settings, and for this demonstration Henry Dale played out the haunts-and-consequences paradigm.

Haunts-and-consequences chronicles the postnatal life course of a natural pattern. Each pattern begins as an embryo, tucked within the manifold potentials of a pattern-assembly system and constrained only by its "universal assembly laws," the physics within which the embryogenic system must operate. Then, a templet appears—a particular context of pattern germination arises, the pattern-assembly system merges with the templet, and the final pattern is born. Later, during their postnatal lives, natural patterns can be quick, vital, and animate. Some patterns dissolve or dissipate before leaving a significant mark. Other patterns beget progeny and pass on their own particular forms—their configurations—to descendant patterns; these parent patterns act as templets for their children. Physiology's *haunts-and-consequences* chronicles the life of a pattern in its role as the templet for succeeding patterns. The physiological, postnatal life-chronicles of a templet has two components. First, there is location: What are the normal haunts of a templet? Second, there are activities: What are the normal consequences of the templet?

To begin, the physiologist chooses a biological pattern to study in its role as a templet. Henry Dale chose to chronicle the actions of the molecule acetylcholine. How did acetylcholine first capture Dale's interest?

> About 1913, it was an accidental observation of the unusual activity of a particular extract of ergot, which quickened anew my interest in phenomena suggesting a chemical, pharmacodynamic transmission of excitation at the junctional contacts between nerve-endings and cells. What was supposed to be an ordinary liquid extract of ergot had been sent to me for a routine control

of its activity. When a conventional dose of this was injected into the vein of an anaesthetized cat it caused a profound inhibition of the heart-beat...Tests of this extract on other biological reagents, such as isolated loops of intestine, confirmed the presence in it of an unusual constituent, with actions suggestively resembling those of muscarine...but, when it...was tested, it became clear that it could not be the stable muscarine; it was something with the properties, rather, of a very labile ester. There seemed little hope then of further progress with the minute amount [of this substance] in hand, till I recalled into conscious memory an observation made some 8 years earlier by my late friend Dr. Reid Hunt [who] had found that, when choline was acetylated, its depressor activity was enormously intensified, the unstable ester, acetylcholine, being some ten thousand times as active as the parent choline. When [A.J.] Ewins, accordingly, made me some acetylcholine, its identity with our substance from the ergot extract was immediately put beyond doubt...

[In 1929], after an interval of some 15 years [and] in a new atmosphere of generally awakened interest in its possible significance [notably through Otto Loewi's reaearch], my direct participation in experiments with acetylcholine was renewed, and this was stimulated, again, by another accidental encounter with the substance. As a late item in a series of studies on the distribution of histamine...we found that extracts from [spleens] also contained something having the activity and the chemical properties of acetylcholine, and were thus led to the first chemical isolation of that ester from an animal tissue. And, having acetylcholine thrust, as it were, again upon my notice, I was led to embark, with a succession of good colleagues, on studies of the wider physiological significance of its activities, as a transmitter of nervous effects.

[Dale, 1953, pp. xi and xiv-xv]

Eventually, Dale purified acetylcholine from horse spleens, crystallizing out the molecule as a platinum salt; from 33 kg of spleens, Dale produced 0.000064 kg of pure acetylcholine crystals. To carry out the haunts-and-consequences paradigm, one must be able to recognize his biological pattern in situ and to follow it during its natural travels. Dale's isolation and purification procedures of acetylcholine were an acceptable vital stain for the molecule, but they involved time-consuming methods that were not practical for microscopic analyses, for experiments necessitating many separate analysis, or for observations directly in situ.

In 1933, Wilhelm Feldberg changed this situation when he brought a simple, specific, and sensitive assay for acetylcholine into Dale's laboratory. First, the biological preparation (for example, an isolated muscle or organ of an anesthetized cat) was stabilized by the addition of physostigmine.* Next, blood or perfusates from the preparation were applied to the back muscles of a leech (*Hirudo officinalis*), which contract strongly in response to small amounts of acetylcholine. Finally, curare was added to the leech muscle; curare is a reversible competitive neuromuscular blocking agent, and curare's abolition or prevention of the contraction documented that the response had actually been due to acetylcholine. The complete assay was a reliable and efficient vital stain for acetylcholine, and it permitted the rapid progress of the succeeding physiological experiments on chemical synaptic transmission.

For acetylcholine, Dale's legacy is a thorough demonstration that this chemical is the mammalian neuromuscular transmitter. Along the way, Dale also delineated many other natural haunts of the acetylcholine molecule. Dale and Dudley's purification of acetylcholine from horse spleens was the first proof that acetylcholine is present naturally in animal tissues in significant quantities. Dale and Feldberg found acetylcholine in venous blood and in perfusates of mammalian stomach walls; then they demonstrated that, when the vagus nerve innervating the stomach was stimulated sufficiently to produce stomach contractions, the amount of acetylcholine increased as much as fourfold. Dale and Feldberg also showed that acetylcholine appears in mammalian sweat glands when the incoming autonomic

---

*Physostigmine—eserine—is an alkaloid from the Calabar bean of tropical West Africa, and it inhibits the normally rapid destruction of acetylcholine by the natural acetylcholinesterases of the body.

nerves are stimulated. Finally, in 1936, Dale, Feldberg, and Vogt proved that acetylcholine is released by somatic nerves at the mammalian neuromuscular junction. By the early 1940s, Dale and his co-workers had shown that the natural haunts of acetylcholine included many different synaptic junctions throughout the peripheral nervous system.

In addition to identifying the locations of a templet, the physiologist wants to chronicle its action—this means linking it reproducibly with certain consequences. Can the templet be associated invariably with the appearance of particular new patterns? Here, the physiologist's goal is to document that certain progeny patterns are natural consequences of the templet, and his proof is the demonstation that the templet is a natural and sufficient predecessor for the new patterns. For synaptic transmission, acetylcholine was a possible templet, and post-synaptic responses were candidates for progeny patterns; as of 1914, one could hypothesize that synaptic transmission was a special event mediated by a few specific chemicals, such as acetylcholine. After he verified that its natural haunts included the expected synaptic locations, Dale documented the consequences of acetylcholine. In 1936, Brown, Dale, and Feldberg then applied acetylcholine directly to mammalian skeletal muscles (the gastrocnemius muscles of the cat), and they found that:

(a) Modest doses of acetylcholine always caused muscular contractions with the expected time-course and of the appropriate tensions.
(b) This response was abolished by curare, just as is the natural response to skeletal nerve stimulation.
(c) The response was enhanced by physostigmine, just as is the natural response to skeletal nerve stimulation. Moreover, the physostigmine enhancement was abolished by curare.

VI

> The causal laws in a specific problem cannot be known *a priori*; they must be *found* in nature.
>
> [Bohm, p. 4]

Dale's particular version of the haunts-and-consequences paradigm has been transcribed as a set of tests that can be used to identify neurotransmitters in general. For a compound to be a neurotransmitter, two classes of observations must be demonstrated:

1. Under natural circumstances, the compound must be
   a. present in the appropriate presynaptic terminals
   b. released by presynaptic nerve stimulation.
2. When exogeneously applied, the compound must
   a. mimic the natural postsynaptic responses to nerve stimulation
   b. show the same chemical sensitivities as natural nerve stimulation.

The first two criteria delineate the compound's haunts; the second two criteria document its consequences. For acetylcholine, Dale showed that the compound is present normally in neuromuscular junctions and that it is released by presynaptic stimulation; he also demonstrated that injections of acetylcholine can mimic the natural response to nerve stimulation and that both the natural and the artifical responses are sensitive similarly to curare and to physostigmine. Therefore, acetylcholine fulfills all four of Dale's tests.

At first glance, Dale's tests seem redundant and a bit roundabout. If acetylcholine is the neuromuscular transmitter, should not a single criterion be sufficient? How many criteria are really needed to prove that something is a transmitter? To answer, consider a simpler situation: suppose that we would like to discover the stimulus for daylight. What is the natural "transmitter" of morning? One possibility is that the regular rising of the sun evokes morning. We might, however, entertain other possibilities—the famous English mathematician George Boole recounted that:

> A little friend of the author's, on being put to bed, was heard to ask his brother the pertinent question—"Why does going to sleep at night make it light in the morning?" The brother, who was a year older, was able to reply, that it would be light in the morning even if little boys did not go to sleep at night.
>
> [Boole, p. 361]

The sun rising in the morning is the required predecessor for daylight, a little boy's awakening is not. The invariable association of sunrise and daylight satisfies us that sunrise is the true "transmitter" of morning; on the other hand, the inconstant association of children's rising with daylight convinces us that awakening does not cause morning.

The standard stimulus-and-effect relationship depends on special temporal linkages, and these linkages come in distinct varieties. First, there is *natural precedence:*

> If every occurrence of *B* is predictably preceded by an occurrence of *A*, then *A* is a natural predecessor of *B*.

Every time we find a *B*—a daylight, for instance—we will always have found an A—a sunrise. A second type of temporal linkage is *sufficient precedence:*

> If every occurrence of *A* is inevitably followed by an occurrence of *B*, then *A* is a sufficient predecessor of *B*.

Every time that we find an *A*—a sunrise, for instance —we will always find a *B*—a daylight. Sunrise is both a natural and a sufficient predecessor of daylight; children's awakening is neither a natural nor a sufficient precedent for morning.

Dale's criteria include a number of steps, because they test both natural and sufficient precedence. The test for the natural precedence of a neurotransmitter is:

> If the postsynaptic response *B* is always preceded by the appearance of the compound *A* in the synapse, then *A* is a natural neurotransmitter.

This is Dale's first test: nerve stimulation must be associated invariably with the appearance of the candidate transmitter, and natural precedence is the "haunts" side of the haunts-and-consequences paradigm. The test for sufficiency is:

> If the appearance of the compound *A* in the synapse is inevitably followed by the postsynaptic response *B*, then *A* is a sufficient neurotransmitter.

This is Dale's second test: the application of the candidate transmitter must be followed invariably by an appropriate postsynaptic response, and sufficiency is the "consequences" side of the haunts-and-consequences paradigm.

In logic, causes are usually classed as *necessary* and/or sufficient, not as *natural* and/or sufficient—but in actuality, causes are rarely necessary. Necessity is an absolute and exhaustive concept; and, because of the incondensable complexities of the natural world, we cannot always deal in absolute and exhaustive causes. For example, acetylcholine is the natural and sufficient transmitter at the mammalian neuromuscular junction: release of acetylcholine is a natural and sufficient cause for postsynaptic excitation. However, acetylcholine is not necessary, because other compounds can also be causes of postsynaptic muscular stimulation. Methacholine and carbachol are two good neuromuscular agonists that have been invented by humans, and nicotine (an alkaloid found in the tobacco plant *Nicotiana tabacum)* and arecoline (an alkaloid of the betel nut *Aredca catechu,* which is a euphoretic chewed in India) are two good neuromuscular agonists that have been invented by Nature. Moreover, Nature has not been limited to choline esters or to alkaloids when fabricating functional somatic neuromuscular synapses: the amino acid glutamate is the natural and sufficient excitatory transmitter at arthropod neuromuscular junctions. In the natural realm templets *are* because they *are.* The templets at work in the real world need not be the necessary causes, the best causes, the most efficient causes, the simplest causes, or the only causes—templets are just natural and sufficient predecessors.

VII

The physiological tradition now permeates all of science; machine understandings—machine-based definitions—have become the standards for any analysis. When a concept is phrased in terms of man-made devices, we can describe all the inner parts and their interactions, and the descriptions can be translated into operations that we can actually carry out—and, in the end, we understand best that which we can do. The algorithm is a machine-based definition; for the haunts-and-consequences paradigm, the general algorthm is:

THE HAUNTS—AND—CONSEQUENCES ALGORITHM

MAIN PROGRAM

Identify a templet
- **Choose a pattern to study as a templet**
- **Find a vital stain for the pattern**

Delineate the templet's haunts
- **Vitally stain the templet**
- **Search it out in natural settings**

Document the templet's consequences
- **Exogenously apply the templet or otherwise control its appearance**
- **Record the temporal correlation of the presence of the templet with the appearance of a new progeny pattern**

PATTERN PERCEPTIONS

Chapter Nine

FRAMES OF REFERENCE

Abstractions, such as the three operational paradigms of Pattern Biology, are artificial; they are limited human constructs, tools that help us order and organize our scientific activities. To use them best, we should look at our abstractions through the world and not look at the world through our abstractions. Begin with Nature and her particular phenomena. Each natural phenomenon, each real world example, is a crystal—a monadic perspective, a particular frame of reference—that helps us to understand the appropriate edges, boundaries, and limits of our abstractions. The abstractions do not tell us the edges, boundaries, and the limits of the world—the world is incondensably complex: it has no edges, boundaries, or limits.

Anthropos apteros for days
Walked whistling round and round the Maze,
Relying happily upon
His temperament for getting on.

The hundredth time he sighted, though,
A bush he left an hour ago.
He halted where four alleys crossed,
And recognised that he was lost.

. . .

Anthropos apteros, perplexed
To know which turning to take next,
Looked up and wished he were the bird
To whom such doubts must seem absurd.

[W.H. Auden]

I

Nature's patterns are tangled and interwoven. Nonetheless, although they interdigitate, distinct natural patterns can be delineated; and it is not only humans who identify natural patterns midst the swirl and the jumble of the universe's myriad interacting elements—Nature also distinguishes certain patterns, and for this she uses independent matching, the same criterion used by humans for discerning patterns. A pattern is always identified by an observer. Operationally, an observer is a matching entity, and two patterns are distinct if each is recognized by a different matching entity. For instance, the Maze of Auden's poem is observed by two different matchers; thus, Auden's Maze comprises two distinct patterns:

(a) One pattern is the global two-dimensional geometry of the Maze; this pattern is the planar curves formed by the edges of the walls as seen from above, and it is identified by an aerial observer, the bird.

(b) The other pattern is the local three-dimensional geometry of the Maze; this pattern is the intersecting planes formed by the full walls as seen from the sides, and it is identified by an earthbound observer, the wingless man *Anthropos apteros.*

Two distinct patterns stem from two independent perspectives. A perspective or a matcher defines a frame of reference—and a well-defined frame of reference is an appropriate level of analysis for any pattern.

II

Naturalists have always arranged organisms in hierarchies. On Aristotle's *scala naturae,* the simpler life forms like sponges and fungi were set at the lowest level, complex animals like frogs and lizards were higher, monkeys and humans were near the top, and divinities sat at the pinnacle. Phylogenetic hierarchies *among* organisms have also been compared to hierarchies *within* organisms—especially within metazoans. Ancient and simple organisms have often been equated with the early embryological stages of multicellular organisms, whereas contemporary and complex organisms have been compared to the adult stages. The German zoologist Ernst Haeckel, for instance, drew a tree of life in the spirit of "ontogeny recapitulates phylogeny" (see Figure 3.1); this tree outlined a direct parallel between an evolutionary hierarchy and a developmental hierarchy. Haeckel began with the newly fertilized egg, and phylogenetically his tree was rooted in *Moneren* (prokaryotes). The lower trunk ascended from *Amoeben* (primitive eukaryotes), which Haeckel compared to advanced fertilized eggs; *Gastraeaden* (spongelike two-layered animals) were mid-trunk level. Next came various embryonic forms, and these were paralleled phylogenetically by the worms. Upper branches included advanced "fetal forms": fish, frogs, and reptiles and then primitive mammals, marsupials, and advanced mammals. Finally, at the crown of the great tree were monkeys, apes, and men—Nature's fully matured progeny.

Biologists also organize their investigations hierarchically: there is Molecular Biology, Cell Biology, Systems Biology, and Organismal Biology. Within each discipline, particular, distinct, and fundamental natural architectures have been identified; these architectures are the biological patterns that have proved useful in making sense of those phenomenena uniquely within the frame of reference of the field. In this way, biological architectures are also studied hierarchically. Molecular Biology attends to the architectures and the interactions of molecules and of metabolic reactions, Cell Biology to the architectures and the interactions of macromolecular complexes and of organelles, Systems Biology to the architectures and the interactions of tissues and of organs, and Organismal biology to the architectures and the interactions among populations of organisms. The architecture of glycoprotein molecules is distinct from the architecture of the linguistic repertoire of monkeys, because these two patterns are composed of different natural units. Nonetheless, biologists attempt to relate distinct architectures found at different hierarchical levels; for example, a biochemist might say to his ethological colleagues: "Perhaps, the naturally occurring patterns of monkey communication derive ultimately from the architectures of glycoprotein molecules in monkey brains."

This (the biochemist's) suggestion comes out of the particular way that reductionism has been applied successfully in Physics. In Physics, higher level architectures have often been well explained in terms of lower level structures. And like the physical architectures themselves, the physical explanations have been hierarchical; for useful explanations, the physicist has always peered downward. But is this the only direction for scientific reductionism? Is scientific explanation always hierarchical, and does reductionism—explanation in terms of different patterns—always proceed towards explanatory patterns at a lower level in the hierarchy? I think not. In Biology, where distinct architectures develop and evolve interactively, the hierarchy can sometimes be turned on its head. In Biology, an evolution-minded ethologist can suggest to the biochemist: "Perhaps, the naturally occurring architectures of brain glycoproteins derive ultimately from linguistic interaction among the animals."

Explanations need not always well up from lower levels of the natural hierarchy. Scientific explanation is a form of reductionism—however, reductionism does not need to make things smaller: scientific reductionism is only translation. As Ernest Nagel [p. 338] wrote:

> Reduction [is simply] the explanation of a theory or a set of experimental laws established in one area of inquiry, by a theory usually though not invariably formulated for some other domain.

The first step in a pattern biologist's reductionism is to take a pattern that he has recognized as a whole gestalt and to translate it into its natural architecture, defining both the natural content and the natural configuration. The natural content and the natural configuration will lie on a plane in the biological hierarchy, and the whole pattern will automatically settle at this, its natural level. The right level for analyzing a particular biological pattern is dictated by its natural architecture, but how are we to discern this natural architecture?

To recognize natural architectures, we must ask Nature. Nature uses interactive elements to identify her units, and in the broadest terms interactive elements are matchers. Two matching entities define and identify each other;* each matcher determines the necessary boundaries, edges, center, extent, and nature of its matching companion. Biological matching entities include enzymes and substrates, retinas and light, hormones and receptors, ears and sounds, olfactory receptor cells and odors, antibodies and antigens, motorneurons and muscle cells, insects and flowers, and parasites and hosts. Natural matching determines natural units—Nature's matchers define Nature's parts—and a matcher is the operational definition for the frame of reference within which some pattern is identified . The observer's frame of reference determines the patterns to which he attends, and the same physical structure (like Auden's Maze) can comprise a number of distinct patterns, because the pattern units can vary according to the frame of reference. To justify our choice of units, we must specify explicitly the observational frame of reference: operationally, we must specify the entities that match those units.

For instance, what does Nature consider to be the natural units of a molecule like insulin? At the protein synthesis level, natural matching is specific for particular amino acids. In terms of protein construction, amino acids are the natural units; ribosomes are the natural synthesizing machinery, and the attendant transfer RNAs are the natural matchers that directly identify particular amino acids. At the same time, natural matchings (and, thus, naturally distinct patterns) occur at all levels in the biological hierarchy. At the hormonal level, membrane receptors are the natural matchers for insulin: insulin receptors identify certain natural functional units of the insulin molecule. Here, the three-dimensional structure of the folded polypeptide chain embodies the natural units, and overall molecular characteristics, such as global configuration, charge regions, and hydrophobic segments, become critical. From this frame of reference, the amino acid subunits take on unequal importance. (It appears, for instance, that four amino acids in the A chain and five amino acids in the B chain are especially determinative.) And at this level—in contrast to the translational level—natural matching is insensitive to certain amino acids. Cow

*"To identify" means "to match": "identify" comes from the Latin word *idem* which means "the same" or "matching."

insulin and pig insulin differ from human insulin in from one to three of the approximately fifty amino acids, but these animal hormones substitute quite well for the human hormone, and they have been used commonly as treatments for human diabetes.

The matchers (the hormone receptors) at the functional level have a perception that differs from the matchers (the transfer RNA's) at the protein synthetic level, and neither level is absolutely right or universally better for all analyses. All levels of the biological hierarchy are important, and each should be thoroughly explored; as the bacteriologist Rene Dubos [pp. 337-338] explained:

> In the most common and probably the most important phenomena of life, the constituent parts are so interdependent that they lose their character, and their meaning and indeed their very existence, when dissected from the functional whole. In order to deal with problems of organized complexity, it is therefore essential to investigate [directly those levels in the natural hierarchy] in which several interrelated systems function in an integrated manner.

And the chemist Michael Polanyi [p. 1312] has written:

> The idea, which comes to us from Galileo and Gassendi, that all manner of things must ultimately be understood in terms of [atomic] matter in motion [can be] refuted [because the] spectacle of physical matter forming the basic tangible ground of the universe is found to be almost empty of meaning [by itself]. The universal topography of atomic particles...which, according to Laplace, offers us a universal knowledge of all things is seen to contain hardly any knowledge that is of interest.
>
> The claims made, following the discovery of DNA, to the effect that all study of life could be reduced eventually to molecular biology, have shown once more that the Laplacean idea of universal knowledge is still the theoretical ideal of the natural sciences...But [an] analysis of the hierarchy of living things shows that to reduce this hierarchy to ultimate particulars is to wipe out our very sight

> of it. Such [an] analysis proves this ideal to be both false and destructive. Each separate level of [life is] interesting in itself and can be studied in itself.

Study is equally important at all levels, and reduction to a lower level is not automatically necessary for explanations. Reduction—translation—to other patterns via other frames of reference is critical to our understanding, but useful pattern analyses can transpire within levels of the biological hierarchy as well as between levels.

III

Decomposition-with-Bookkeeping

The pattern biological paradigms have been useful at all levels of the biological hierarchy; as examples, let me briefly survey the Nobel Prize-winning work in Biology. First, consider *decomposition-with-bookkeeping,* the definitional operations that discover the content and the configuration of a biological pattern. At the level of molecular architecture, *decomposition-with-bookkeeping* is exemplified by Frederick Sanger's sequencing of insulin. In a similar way, Robert Holley worked out the complete content and configuration (the one-dimensional sequence) of the alanine transfer RNA molecule, Sune Bergstrom and Bengt Samuelsson purified and discovered the full content and configuration of a prostaglandin, Christian Anfinsen, Stanford Moore, and William Stein unpuzzled the complete content and configuration of the enzyme ribonuclease, and Roger Guillemin and Andrew Schally purified, determined the full content and configuration of, and synthesized three neural control polypeptides: thyrotrophin releasing factor, gonadotrophic releasing factor, and somatotrophic release inhibiting factor (somatostatin).

Natural chemical reactions have particular forms—natural architectures—in situ. Hans Krebs inferred the full content and configuration of a ubiquitous sequence of intracellular chemical reactions by which energy is produced; this sequence is the tricarboxylic acid cycle, the central pathways of intermediary metabolism. Konrad Bloch and Feodor Lynen identified the constituents and the sequence (the content and the configuration) of the metabolic pathway by which cholesterol (the precursor molecule for all the steroid hormones) is naturally synthesized; and the metabolic pathway by which cholesterol is catabolized has been unpuzzled by Michael Brown and Joseph Goldstein, following their discovery of the LDL receptor, which mediates the specific cellular uptake of cholesterol.

Animal behaviors also form distinct biological patterns: behaviors are architectures at the psychological level. Karl von Frisch, Konrad Lorenz, and Nikolaas Tinbergen developed the concept that both learned and instinctive animal behavior can be analyzed into sets of smaller behavioral units—the behavioral content; these units are then played out in a particular stereotyped program—the behavioral configuration. Roger Sperry began a unique mapping of the content and the configuration of higher brain functions. For content, Sperry showed that each cerebral hemisphere represents a separate and independent functional unit, with its own special major ability, either linguistic or spatial; for configuration, he showed that the relevant interconnection between these units is largely via the corpus callosum.

Often, *decomposition-with-bookkeeping* is applied in separate parts, such as the purification subroutine—a set of operations by which natural units are produced in homogeneous populations of sufficient bulk and of sufficient uniformity to be studied in the laboratory. To define a molecule, one needs a pure preparation—Theodor Svedberg perfected centrifugal methods for purifying compounds by their molecular size and weight, and Arne Tiselius devised electrophoretic separation of molecules, especially proteins. Later, John Northrop, Wendell Stanley, and James Sumner purified and crystallized a number of enzymes and showed convincingly that homogeneous enzymes are stereotyped molecules of particular proteins. The first biological molecules to be purified are used subsequently as models for understanding whole categories of related molecules; among the archetypic biological molecules were: the tobacco mosaic virus, crystallized by Northrop and Stanley; cholorphyll and a number of carotenes, purified and chemically studied by Richard Willstatter; and penicillin, which was purified by Ernst Chain and Howard Florey. Another good technique for purifying is direct synthesis. Albert von Szent-Gyorgyi showed that vitamin C is ascorbic acid, which could be purified as crystals; and Walter Haworth described the structure of ascorbic acid and then synthesized it. Tadeus Reichstein first purified and synthesized adrenocortical steroids, and Vincent de Vigneaud synthesized the first polypeoptide hormone, oxytocin. More recently, Georges Kohler and Cesar Milstein developed the techniques for producing large quantities of pure (monoclonal) antibodies with precisely designed specificities.

Metabolic reactions can also be purified: Eduard Buchner purified the fermentation reaction, producing cell-free alcoholic fermentation in vitro from cell extracts. Macromolecular complexes have been purified: Albert Claude, Christian de Duve, and George Palade developed and applied a variety of subcellular fractionation techniques, especially using differential centrifugation, to purify specific intracellular architectures, such as organelles. Organisms, too, can purified. Microbiology is founded on the purification methods developed by Robert Koch, who devised today's gelatin and agar techniques for growing homogeneous bacterial cultures. (The quarter century following Koch's work has been called the "Golden Age of Bacteriology," because during those years most of the major bacterial pathogens were isolated and described using Koch's techniques.) Later, Selman Waksman methodically repeated the purification subroutine with the actinomycetous bacteria and eventually isolated pure colonies of *Streptomyces griseus* that produce the antibiotic streptomycin. (Streptomycin was the first antibiotic effective against tuberculosis.) For viruses, André Lwoff worked out methods for regularly producing quantities of bacteriophages, the viruses that infect bacteria. John Enders, Frederick Robbins, and Thomas Weller then developed tissue culture methods for growing poliomyelitis viruses, and this set the stage for the production of antipolio vaccines.

After purifying, pattern biologists discover content. Biochemists struggled for many years to find rules and regularities in the architectures of large organic molecules; Axel Hugo Theorell took a critical step in making sense of enzyme architectures when he discovered that the content of a number of key oxidizing enzymes (notably, NAD and cytochrome c) comprise two units: a protein moiety, and a coenzyme (built of chemical groups that do not form the usual amino acid side chains). For nucleic acids, Albrecht Kossel discovered that the base content of DNA is always: cytosine, thymine, adenine, and guanine. The content of an antibody molecule (Figure 9.1) differs in different natural frames of reference. Rodney Porter identified three structurally significant units, two alike units (the Fab fragments) and one different unit (the Fc fragment); these are natural posttranslational units. Gerald Edelman identified two functionally significant units,

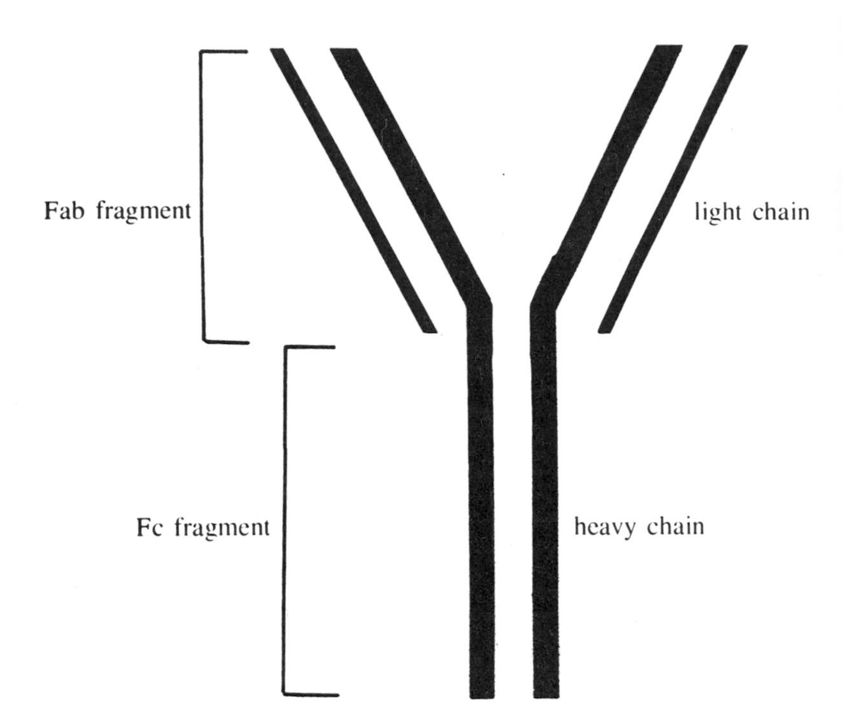

Figure 9.1

Schematic diagram of an antibody molecule. An antibody molecule embodies more than one natural pattern, each with its own set of natural units. One set of natural units—synthetic or translational units—are: light chains and heavy chains. Another set of natural units—postsynthetic or posttranslational units—are: Fab fragments and Fc fragments.

light chains and heavy chains; these are natural translational units.

In situ, metabolic reactions form reproducible biological patterns, each with a particular and stereotyped content. For the metabolic pathways burning amino acids, carbohydrates, and fats, Fritz Lipmann discovered that the compound acetyl coenzyme A is a ubiquitous obligatory intermediate. Carl and Gerty Cori studied the chemical reactions converting glycogen into the activated form of glucose (glucose-6-phosphate) and the obverse reactions converting the activated form of glucose into glycogen; then the Coris discovered that the contents of both reactions include a transient intermediate compound, glucose-1-phosphate, as well as the enzyme phosphoglucomutase. Julius Axelrod discovered that the enzyme catechol-O-methyltransferase is a constituent of the metabolic pathway by which the neurotransmitter norepinepherine is inactivated, and Earl Sutherland Jr. discovered that the molecule cyclic AMP is a common intermediary (a "second messenger") intervening between many hormones and their intracellular actions.

The first third of this century saw the discovery of the content of the metabolic pathways of energy production; in the second third of the century, biologists discovered the content of the metabolic pathways of protein production—the metabolic routes to and from the genes. Along these routes, polynucleotides are synthesized, and so are proteins. Severo Ochoa found that one metabolic pathway by which RNA can be synthesized from ribonucleotides includes the enzyme polynucleotide phosphorylase, which he purified. Arthur Kornberg discovered the DNA polymerases in the metabolic pathway through which DNA is synthesized from deoxyribonucleotides. These metabolic pathways, through which genes are turned into proteins, are not always ballistic—often, the transcription-translation route is cybernetic, and it is regulated by feedback mechanisms. Francois Jacob and Jacques Monod discovered that the content of the transcription-translation pathway includes two "invisible" classes of units that are critical to a genomic feedback mechanism: regulatory genes, and the protein products of those genes.

At the level of subcellular units, Christian de Duve discovered that the lysosome—a membranous organelle filled with destructive enzymes—is a constituent unit in the intracellular pattern that forms the degradative pathways for many different molecules. At the level of tissue patterns, Charles Sherrington discovered that the content of peripheral "motor" nerves includes two types of axons: half of the axons innervating a skeletal muscle are sensory (afferent), and half are motor (efferent). Henry Dale proved that parasympathetic and skeletal motor axons secrete acetylcholine, and Ulf von Euler discovered that sympathetic axons contain norepinephrine. At the level of organismal population patterns, Karl Landsteiner discovered that humans can be divided into four major blood types: A, B, AB, and O.

In addition to determining content, *decomposition-with-bookkeeping* also defines configuration, the architecture via which the units of a pattern are woven into some particular natural entanglement. Walter Haworth determined the architectures of glucose and of a variety of other more complex carbohydrate molecules. Using X-ray diffraction techniques, Maurice Wilkins determined the helical arrangement of the polynucleotide chains of DNA molecules, and Aaron Klug worked out the detailed configurations of a variety of viruses, proteins, and nucleosomes. With scale models, Francis Crick and James Watson inferred the double-helix base-paired configuration of DNA molecules.

Biological molecules like these are tiny structures, and the architectures of such minute objects can be exposed with the *equivalence class subroutine* of the decomposition-with-bookkeeping paradigm. Equivalence class analysis requires careful fracturing of the objects of study; thus, the invention of precise scalpels has always stimulated new configurational analyses. The fracturing methodologies devised by Stanford Moore and William Stein led to the automated amino acid analyzer for proteins. Similarly, organizational analysis of the genome was made into a common laboratory operation when Werner Arber, Daniel Nathans, and Hamilton Smith discovered the restriction endonucleases, bacterial enzymes that cut DNA strands into pieces. Using these and other enzymes, Walter Gilbert and Frederick Sanger developed fast and accurate technologies for fully sequencing DNA molecules.*

---

*Genetics takes a less direct approach to unpuzzling the configuration of the genome. Using genetic phenotype analysis early in this century, Thoman Morgan discovered that the genome has a fairly simple linear configuration: genes are arranged in long

At the level of metabolic reactions, Albert von Szent-Gyorgyi worked out the functional configuration (the order) of enzymatic reactions in cellular respiration, and Melvin Calvin inferred the full sequence of metabolic reactions for photosynthesis. At the level of functional tissue architectures, Charles Sherrington set the form of contemporary neurophysiological studies with his demonstration of the orderly topographical maps of motor information in both voluntary and reflex areas of the vertebrate central nervous system; Sherrington saw the complex operation of the brain as manifold interacting patterns, where these patterns are physical maps of reflex actions playing out across the intricate architectures woven into the populations of neurons.**

In addition, biological patterns have configurations that are organized in the temporal dimension as well as in the spatial dimensions. For populations of organisms, the temporal sequence of natural events was originally mapped by matching fossils and other organic relics directly to the geological strata in which they were found; today,

---

sequential strings. Later, Joshua Lederberg extended this result to bacteria—demonstrating that the configuration of a bacterial genome can be analyzed in the same way as the configuration of a metazoan genome, such as the genome of ***Drosophila.***

**Sherrington [pp. 174, 176, and 178] wrote:

> Of all people the reacting individual himself is the last to think himself reflex, [but] such a reflex system, operated by sequences of thousand-patterned stimuli [that stem from the full] situations of the moment, might well work a Robot for many purposes indistinguishable from a man.
>
> Imagine [that our brain] activity [were shown by] little points of light. [Our behavior would then appear] as if the Milky Way entered upon some cosmic dance. Swiftly the [brain would] become an enchanted loom where millions of flashing shuttles weave a dissolving pattern, always a meaningful pattern though never an abiding one; a shifting harmony of subpatterns...Dissolving pattern after dissolving pattern will, the long day through, without remission melt into and succeed each other.

radiocarbon dating—developed by Willard Libby—provides an independent method for placing organic remains within the overall sequence of natural history and thereby defines temporal architectures.

Alternative Architectures

Alternative architectures is a set of explanational operations with two tasks:

(a) to uncover the full range of configurations inherently available to the given biological system, and
(b) to delimit and model the templets that actually operate in a particular situation.

The genes of an organism form a genetic architecture, and, using X-rays, Hermann Muller demonstrated that the possible genetic architectures are more wide-ranging than is apparent from studying common organisms. Moreover, the classical picture (proposed by Thomas Morgan and his colleagues) had presumed an inherently fixed genetic order, but Barbara McClintock showed that genetic architecture is inherently malleable: some genes—transposable genetic elements—can autonomously change their order, further increasing the range of possible genetic architectures.

Another critical molecular architecture is formed by the ionic patterns across cell membranes, especially in the nervous system. For instance, the ionic patterns (the generator potentials) produced in the sensory terminal of a neuron initiate the transmission of signals by the cell. Bernard Katz showed that a generator potential is proportional to the strength of the stimulus and that generator potentials can vary over a wide range. In other words, the intrinsic properties of a sensory terminal can produce all manner of depolarizations, the range of alternative ionic architectures is broad, and the final particular output pattern is not constrained to a few inherent and predetermined configurations; instead, the output of a neuron is extensively templeted by its input. Ionic patterns are architectures of simple chemicals, whereas metabolic reactions form architectures of complex chemicals. Frequently, such biochemical reactions appear naturally with only one of a number of available configurations—for instance, many metabolic reactions that are quite reversible in vitro operate unidirectionally in situ. (The polynucleotide phosphorylase that was discovered by Severo Ochoa forms RNA-like polymers from ribonucleotides; the reaction does not need a DNA template, and in vitro it is readily reversed by increasing the concentration of endproducts.)

At the cellular level, the configurational organization of a set of organelles can influence cell function. Bernard Katz showed that neural synaptic transmission is the coordinated release of a horde of tiny organelles, vesicles containing neurotransmitter. Katz then demonstrated that in the unstimulated system these packets are normally released independently, stochastically, and without coordination. Inherently, quantal packets of neurotransmitter can be released in a great variety of configurations, and the presynaptic potentials act as templets that organize a coordinated pattern—a particular architecture—of massive release.

Tissue level architectures can also have a wide range. In the circulatory system, the capillary beds form configurations defined by diverse and complex states of contraction and distension. August Krogh found that blood capillaries can independently assume all combinations of these states—that is, the range of alternative capillary architectures is large. Moreover, Krogh discovered that blood circulation is templeted at the level of the capillaries and not at the level of the feeder arterioles. Another good example is the immune system; this tissue forms complex biochemical patterns that include an "intrinsic" fingerprint of self—a self-recognition immune pattern. Frank Burnet and Peter Medawar applied the alternative architectures paradigm and varied the developmental circumstances for a set of mammals; in this way, they discovered that the self-recognition patterns of an organism are (in part) acquired properties. These patterns are not completely predetermined genetically, and the range of alternative self-recognition architectures is quite broad. Likewise the range of alternative *neural* architectures is extensive. David Hubel and Torsten Wiesel showed that, during its embryogenesis, the neocortex of the brain can form a variety of different connection patterns, and the final configuration of functional connectivity is templeted by visual experience during critical periods of early postnatal development.

Functionally, the various neural architectures appear as psychological architectures; therefore, genetically preprogrammed (self-assembling) neural circuitry will manifest as limitations on the available alternative psychological architectures. Haldan Hartline found that the neural connectivity genetically built into the eye produces a local lateral inhibition, and this lateral inhibition naturally enhances edges of visual images. Here, the psychological architectures have an inherently limited range: lateral inhibition is preprogrammed into the eye, and it constrains and organizes all the neural output of the visual system.

The range of alternative architectures available from a given system indicates the range of effective templets that Nature can use. When there are many alternative architectures, the range of determinative templets is correspondingly large; in contrast, when the self-assembling properties of the system dominate and the alternative architectures are minimal, outside templets will have a much more modest role. With the alternative architectures paradigm, a pattern scientist first determines the range of potential alternative architectures. Next, he infers the characteristics and the scope of the operant templets. Finally, he tests both his matching matrix (the summary of the alternative architectures) and his inferred templets by modeling the templets and by applying his model to the real world.

At the molecular level, Paul Muller tested the proposal that compounds of the form C1-x-C1 act as templets that produce the consequence "insect death," and from this blueprint Muller built the effective insect poison C1-⟨0⟩-CHCC1$_3$-⟨0⟩-C1, commonly known as DDT. Similarly, to test the proposed details of the genetic code, H. Gobind Khorana built a wide range of polynucleotide templets. Daniel Bovet championed the idea that biochemical *architecture* is *function* in organisms. Under this dictum, synthetic molecules designed to mimic the architectures of natural molecules should also mimic the functions of the natural molecules; thus, Bovet built synthetic structural analogs in order to test the functional import of particular molecular architectures. (Among the useful synthetic molecules that Bovet introduced or studied in detail are: succinylcholine, gallamine (Flaxedil™), dimethylaminoethylbenzylaniline (Anergan™), and the dimethyl amide of lysergic acid, LSD.)

The amounts and varieties of proteins produced by an organism form a particular pattern of genomic output, an architecture of protein production. From their experiments on the regulation of enzymes in bacteria, Francois Jacob and Jacques Monod found that the pattern of genomic output is quite malleable. Extrinsic (extra-genomal) templets can in some cases control enzyme production at the level of transcription, and Jacob and Monod suggested the following model:

> The genome contains two major classes of genes: structural genes and regulatory genes. Structural genes follow the "one gene—one protein" rule. In contrast, regulatory genes are a step removed from the direct protein codes and they may show a "one gene—many proteins" effect. The extrinsic regulatory templets (often, the final protein products that are feeding-back onto the genome) directly control the regulatory genes.

Jacob and Monod's model proposed that individual regulatory templets can have simultaneous effects on the production of more than one different protein, and subsequent tests have confirmed the details of this model.

At the tissue level, George von Bekesy delimited the role of templets in the physics of hearing. The cochlea of the inner ear discriminates sounds by frequency, and von Bekesy proposed that in the cochlea the frequency spectra of sounds are transduced into spatial maps along the basilar membrane. Although sound waves of all frequencies are propagated as traveling waves running throughout the basilar membrane, sharply different spatial patterns are produced by sound waves of different frequencies; thus, von Bekesy inferred that the boundaries (the shape delineators) of the basilar membrane must mechanically constrain the inherently wide-ranging wave effects. In other words:

> (a) The inherent properties of the basilar membrane permit a wide variety of pattern architectures. (These architectures are frequency-to-space transductions.)

(b) A specific templet (the shape of the basilar membrane) is responsible for the specific architecture (the particular transduction) that actually occurs.

After inferring the templet, von Bekesy tested his idea with a direct model. For the mechanical components, he re-created the inner ear as a water-filled plastic tube shaped like the basilar membrane; and for the neural receptor component, he laid his arm along the plastic tube. As he had predicted, von Bekesy felt distinct and different spatial patterns for different sounds along his arm.

Like the nervous system, another tissue that functions through precisely interacting patterns is the immune system. Although it is physically diffuse, the immune system embodies a specific recognition pattern that precisely mirrors its antigenic world; in particular, the immune tissue forms a positive image of an animal's extrinsic antigenic world and a negative image of an animal's intrinsic antigenic world. Neither image is genetically preprogrammed into the organism: the particular recognition architecture of each mammal is unique and idiosyncratic, and it is templeted by the experience of the immune tissue as it develops and as it contacts the external environment. Niels Jerne modelled the creation of this recognition pattern as templeted selection. During development, proposed Jerne, immune cells are culled from a large and diverse preexisting pool of potential antibody-production cells. The final population of the immune system is then a pattern that has been templeted by the survival of only those clones of immune cells that best match the antigenic worlds, and this process is called "clonal selection."

Haunts-and-Consequences

Haunts-and-consequences comprises those chronicling operations that consider biological patterns to be templets that eventually structure new biological patterns. To follow his templet pattern in situ, the physiologist first needs a vital stain. Radioactive vital stains can be highly specific, and George de Hevesy devised techniques for using radiolabelled substances as tracers for diverse in situ chemical processes. Metabolic processes can also be followed in their natural haunts by other means. For example, Archibald Hill developed a quantitative vital stain for the enthalpic chemical reactions taking place in an active muscle; Hill's "stain" was the precise in situ measurement of heat production. For cellular electrical patterns, Joseph Erlanger and Herbert Gasser perfected techniques for amplifying and for recording action potentials from living nerve fibers. For tissue electrical patterns, Willem Einthoven invented the electrocardiogram (a vital stain for the natural electrical actvitiy of the heart); and Walter Hess created precisely localized, permanently indwelling brain electrodes, which can stimulate and record electrical patterns from specific regions of the brain in awake and behaving animals. For monitoring the in situ blood flow patterns of the circulatory system, Andre Cournand, Werner Forssmann, and Dickinson Richards developed cardiac catheterization. And, as a vital staining technique for tissue and organ patterns generally, Godfrey Hounsfield and Allan Cormack invented the computerized X-ray (CAT) scanners for noninvasive body tomography.

Once a pattern has been identified as a templet and after a vital stain has been developed, the pattern biologist maps out the natural haunts of the pattern. After the malarial parasite was identified, Ronald Ross showed that the parasite was naturally found in the salivary glands and the stomachs of mosquitoes. Similarly, Charles Nicolle proved that the louse is the natural transmission agent for epidemic typhus *(Ricketsia prowazekii)*, a bacterial disease. Food acting as a templet increases stomach secretions—are stomachs the natural haunts for this templet? To answer this question, Ivan Pavlov first demonstrated that the vagus nerve normally stimulates gastric secretions; then he showed that the vagus nerve responds to food as a templet. Finally, Pavlov found that the natural haunts of this stimulus are the mouth and not the stomach: gastric secretion is reproduced by action at a distance, and it occurs in sham feeding experiments where food in the mouth can never reach the stomach because of an esophogeal diversion.

Independent of discovering a templet's natural haunts, one can document the templet's biological consequences. Edgar Adrian showed that the mean frequency of nerve firing is the direct consequence of the intensity of the sensory stimulus, and he suggested that many neurons communicate using an intensity-to-frequency code. Otto Meyerhof showed that lactic acid accumulation is the consequence of glycogen disappearance in actively contracting muscle. Severo

Ochoa and Marshal Nirenberg tested the intracellular consequences of carefully engineered ribonucleic molecules; they applied well-defined templets (synthetic polynucleotides of known sequences) to the intracellular synthetic machinery and unpuzzled the genetic code. Elie Metchnikoff found that infectious microbes invariably induced the mobilization of defense cells, which then phagocytose the invaders: defense phagocytosis is a natural consequence of pathogenic microorganisms. In medicine, proving that a particular agent will cause a specific disease is a classic example of documenting the templet's consequences. Robert Koch proved that *Bacillus anthracis* is the cause of anthrax and that *Mycobacterium tuberculosis* is the cause of tuberculosis. Theodor Kocher showed that myxedema inevitably follows removal of the thryoid gland. Frederick Banting and John Macleod found that a pancreatic secretion, insulin, lowers blood sugar and reverses the symptoms of diabetes. And Christiaan Eijkman, Henrik Dam, Edward Doisy, Frederick Hopkins, George Minot, William Murphy, Albert von Szent-Gyorgyi, and George Whipple each helped to demonstrate that vitamins will cure nutritional deficiency diseases.

"Consequences," however, is an umbrella relationship; to better understand the interactions of natural patterns, pattern scientists attempt to identify more precise relationships. One subsidiary relationship that is more exact than "consequences" is *natural precedence:*

> If every occurrence of *B* is predictably preceded by an occurrence of *A*, then *A* is a natural predecessor of *B*.

Henry Dale's tests for neurotransmitters included two steps explicitly documenting natural precedence. For microbiology, Robert Koch proposed a parallel set of tests (Koch's Postulates) to distinguish pathogenic microbes from adventitious microbes; like Dale's tests, Koch's first two criteria for pathogenic microbes establish natural precedence:

(1a) The organism is regularly found in lesions of the disease.

(1b) The organism can be grown in pure cultures from lesions.

Likewise, when George Beadle and Edward Tatum proposed the concept of "one gene—one protein," they also set out to demonstrate natural precedence. Beadle and Tatum studied the bread mold *Neurospora crassa* and found that a single nutritional deficiency—in this case, dependence on external biotin—was always associated with the mutation of one particular gene. The metabolic deficiency was a single enzyme defect; thus, a particular gene was shown to be the natural predecessor of one specific protein.

Another relationship that is more exact than "consequences" is *sufficient precedence:*

> If every occurrence of *A* is inevitably followed by an occurrence of *B*, then *A* is a sufficient predecessor of *B*.

In Koch's Postulates, sufficient precedence appears as:

(2a) Innoculation of exogenous organisms produces the disease.

(2b) Such experimentally induced lesions contain the organism.

At the molecular biology level, Beadle and Tatum not only established natural precedence, they also demonstrated sufficient precedence in the relationship "one gene—one protein." For sufficient precedence, Beadle and Tatum showed that the mutation of the gene previously identified as a natural predecessor always led to the same nutritional deficiency; thus, the same gene was also sufficient for the production of the specific enzyme needed to make biotin. Likewise at the molecular biology level, Alfred Hershey showed that DNA molecules are sufficient to carry genetic information: Hershey found that viral

DNA alone—with no accompanying protein—could always initiate the generation of complete viral progeny. At the systems level, Ivan Pavlov demonstrated the sufficiency of the vagus nerve for causing gastric secretions; when he stimulated the cut end of the nerve, he found this invariably induced gastric secretions of the appropriate amounts. Corneille Heymans demonstrated the sufficiency of a particular peripheral receptor area—the carotid sinus and the carotid body of the aorta—for control of blood pressure and respiration of the whole animal. Heymans isolated this receptor area and subjected it to hypotension and to hypertension; in response, the central nervous system always initiated the appropriate whole body pressor reflexes. Then, when he varied the blood gases in the isolated receptor area, Heymans invariably induced appropriate respiratory reflex responses.

In the real world, natural and sufficient precedence do not imply necessary causality, and they do not determine exhaustive or absolute causality. Nature is luxuriant, and she has recourse to many causes. For instance, from in vitro experiments, Carl and Gerty Cori found that the enzyme glycogen phosphorylase was a natural and sufficient predecessor for the synthesis of glycogen from glucose, via the transient intermediary glucose-1-phosphate; however, another metabolic pathway—using the enzyme glucose-1-phosphate uridylyl-transferase—is equally natural and sufficient, and in situ this latter pathway is the more common naturally-occurring route for glycogen synthesis. For any given situation, the relevant "causes" are simply the natural and sufficient predecessors.

IV

The Pattern Biology paradigms have been useful at all natural levels. To understand biological patterns, one need not translate them into a special common plane of the natural hierarchy; specifically, one need not always rewrite them as patterns at lower levels in the natural hierarchy. Scientific reduction is simply translation: it is the reformulation of one set of things in independent terms. The appropriate level of translation is determined by the frame of reference of the questioner. For an understanding of molecular architecture, the natural frame of reference is molecular; for an understanding of behavioral architecture, the natural frame of reference is behavioral.

Each of the three major modes of pattern analysis is reductionist translation. *Definition* translates a whole pattern into its component units and their interconnections. *Explanation* translates the form of a pattern into its two generative parents: the "universal assembly laws," and a templet. *Chronicling* translates an action into a natural templet and its consequences. In all cases, reductionism discerns a set of different parts associated with the pattern; but the new parts need not always reside at a lower level in the natural hierarchy—as Loren Eiseley [p. 202] wrote:

> I have come to suspect that this long descent down the ladder of life [—that is, the continued reduction to lower levels in the biological hierarchy—] beautiful and instructive though it may be, will not lead us to the final secret. In fact I have ceased to believe in the final brew or the ultimate chemical... Somewhere among the seeds and beetle shells and abandoned grasshopper legs I find something that is not accounted for very clearly in the dissections to the ultimate virus of crystal or protein particle. Even if the secret is contained in these things, in other words, I do not think it will yield to the kind of [downward-looking hierarchical analysis that is common in our standard] science.

Useful reductionism does not necessarily move in one direction along the natural hierarchy, and this saves us from an eventual end to science. If there were ultimate elements, then some day all scientists would have to retire and to become craftsmen or poets. However, science is not about discovering the ultimate things; it is not finding out the final inner workings of the atom or tracing the full genetic basis of courtship rituals. Once the inner workings of the atom have been resolved into some set of terms, they must then be recast in another set of terms—physicists will always dream up new subatomic particles and new transcendental forces. Once the ethologist finds a gene that influences courtship dancing in fish, he will hunt for other critical genes, and he will look for interactive population effects that supersede genetic influences. Each scientist continues to peer and to probe, to reformulate and to resynthesize; there is a temporary pleasure in fitting together a large and messy section of the natural jigsaw puzzle, but the urge to continue working is stronger and soon drives us ever on and on.

Reductionism is never-ending—all observations must be continually translated and retranslated. Scientific reductionism is the expression of a peculiarly human itch impelling us continually to move and to do and to pull things apart and to put things together. The need to fiddle, to shape, and to reshape puts only one demand on the objects with which we play: the puzzles, marionettes, building sets, and other playthings of our toyroom must be reducible. "To reduce" means "to translate into other parts," and our old toys must have parts if we are to translate them into new toys. Like reworkable toys, scientific abstractions must be able to be reapportioned and to be rebuilt. To be practical, a scientific abstraction must be mapped to observations in the real world; but to be philosophically useful, a scientific abstraction must also have parts—this is the quality called "philosophical depth."

The best scientific abstractions have parts; parts, however, need not reside at some lower level of the natural hierarchy—the parts must simply be different than the ones in the original abstraction. Parts are delineated by matchers, by defined frames of reference, and new parts are identified by new frames of reference. A deep scientific abstraction can be translated into *other* parts, and that is the essence of science. Science is not the production of some particular thing or the discovery of some specific final answer. Science is just the unending, orderly, organized translation of the variegated patterns of our Maze; as long as we can continue to create new frames of reference, we can continue to do new science.

Chapter Ten

REPETITION & THE ARTISTRY OF NATURAL PATTERNS

The logic that we use to understand patterns is human at base; we organize our abstractions by rules that make human sense. The methodologies that we enact to analyze patterns are human artifices; we operate by recipes that have been devised by people and that fit comfortably in human laboratories. We also take joy in patterns, and this is, of course, a human joy. But can we apply human aesthetics to natural patterns?

The teensy-weensy spider
Went up the waterspout.
Down came the rain
And washed the spider out.

Out came the sun and
Dried up all the rain,
And the teensy-weensy spider
Went up the spout again.

I

What little spring rough lawn I remember, on a summer evening: we were three little children playing *Mother-May-I.* One child is Mother, the all-powerful Queen on a chessboard lawn. "Get to your places!" she shouts; then we call out: "How many? What kind? How many steps do we find?" And Mother answers: "Michael! You can take two baby steps and one umbrella step" or "Michael! You can take one giant step" or "Michael! You can take three bunny steps" or "Michael! You can take two giant steps and two baby steps." "Mother-may-I?" "Yes, you may," she nods.

Mother stands at one edge of the grass, and we begin at the other; slowly and swiftly, saltarello terpsichore, we bounce and dance across the lawn, until one of us can reach out and touch her. It is chess:

in the eighth rank a pawn becomes Queen, the board re-sorts, the pieces line up along the edges, and we begin again. Who will be the new Queen?—the old Mother knows. She chooses her successor; with measured eye she scans the pawns, and at each turn she carefully assigns the moves. Two more giant steps and Michael will be in reach, so it is best to give him only one bunny step and to let Terry have a scissors step *and* a giant step. Ah, now three baby steps for Terry—even one giant step for Michael will still keep him away. Now a baby step for each and...

It was all determined; the steps were measured, and the way was fated. At each rank, the innumerable possibilities—two bunny steps and three baby steps, one scissors step, three giant steps and one umbrella step, a baby step and a scissor step...—these many possible roads were sorted, blocked, organized, and channelled, and only one of the potential itineraries was actually played out. Mother planned the game, and planning constrains the course of events. As any entity rolls through its life in the universe, it wanders through a maze of possible roads, and without planning who can tell its ultimate itinerary? But with planning, the future curriculum vitae—the chronicle of our eventual life travels—is made predictable and secure. Planning transforms the indeterminate haze of tomorrow into a determinate destiny—"Mother-may-I?" "Yes, you may."

On our spring lawn, the future was precise, strong, fixed, and immutable; Mother knew and saw, and she set the final architecture of our travels from edge to edge. All was planned, and planning comforts the methodical bookkeeper within us. At the same time, however, the full human is a complex crowd; we are many beings at once, and we are always torn between the accountant and the dreamer: someone there is that does not like a Plan. Complete planning lays out an automated land where determinacy is Queen; but, although a well-planned future promises to be a secure, comforting, and predictable kingdom, it will also be boring—and the dreamer in us forfends boredom.

The boredom of planning is why chessmasters resign before the execution of their king. I have queen left to consort my king, you have only your sad king. Relentlessly my queen advances; she narrows the safe ranks for your thin man. She inexorably partitions the board into smaller and smaller free zones, ruling with precise orthogonal calipers the flagstone of your castle haven, until, backed behind his throne, your king is swiftly beheaded in the corner. In an endgame with my queen and king against your lonely king, it is all foregone; it is a Greek tragedy that could be played out by a robot. Here, the endgame is fully planned, and it lives up to the mythic and inhuman ideal of determinacy that begins the *International Chess Code's* DEFINITION AND OBJECT: "Chess," announces the *Code,* "[is] a game in the play of which there is no element of chance..."

II

I remember, I remember
  The house where I was born.
The little window where the sun
  Came creeping in at morn.

. . .

I remember, I remember
  The fir-trees dark and high;
I used to think their slender tops
  Were close against the sky:
It was a childish ignorance,
  But now 'tis little joy
To know I'm farther off from heaven
  Than when I was a boy.

[Thomas Hood]

Planning seems to be of the future, but it is really a relic of the past played out in the present. My appointment book for tomorrow morning reads:

7:45 am—Breakfast
8:20 am—Arrive at office
9:00 am—Committee meeting
10:00 am—Lecture

My plans were made yesterday or last week or last month, and my appointment book is my memory. Putting the plans into effect is my daily activity, and I act in the present.

Planning is determinate action—a plan fixes and constrains the itinerary of some entity as it drifts or careens through the world. Plans are memories made animate. My earliest memories are hazy snapshots: they are whole wide scenes, bright pictures of rooms and rainy streets and huge dark houses. I remember one day—it's in a springtime that's rainlit and green-grey with cloud-covered skies. I am standing outside, a child all alone, small on a sidewalk near my home; the trees' leaves are dark and round like green pebbles, and the bark has a silver-grey sheen that's finely etched with lines of brown. The grass is wet. The walks stretch down the streetside forever, and as far as I can see the street has no ending. A large oak tree stands along the misty street alive and watching me; no one else is all around. And, thoughtful and remotely massive, brown brick houses peer from deep within their dark front porches—they loom in rows akin to a giant-stone orchard of temples old and wise, solemn but not forbidding—each mysterious and warm inside.

We moved to a new neighborhood when I was two-and-a-half years old, and these early memories must precede the move. I have been back since; and things have changed—the street now ends (there is a busy intersection at the far corner), maple leaves are on the trees, the grass is no longer wet, the houses are smaller and people live in them. Did I make up that old memory? I am sure that I did not: each time I recall that afternoon, I always see the same picture. To dream that old dream, I close my eyes or just look off distractedly, and a gentle rainy street, green and grey, thick and live, rolls home to me; it reappears vague, gentle, and secure somewhere. I have remembered it thousands of times, and its memory—its re-creation—is as certain to me as the sunrise. As my mind drifts or careens through the world, one thing in its itinerary can be planned—my childhood street memory; it is a bit of determinate fully-fated future. As Loren Eiseley [pp. 207-208] wrote:

> I suppose that in the forty-five years of my existence every atom, every molecule that composes me has changed its position or danced away and beyond to become part of other things. New molecules have come from the grass and bodies of animals to be part of me a little while, yet in this spinning, light and airy as a midge swarm in a shaft of sunlight, my memories hold, and a

> loved face of twenty years ago is before me still..
> Nor is that face, nor all my years, caught cellularly
> as in some cold precise photographic pattern, some
> gross, mechanical reproduction of the past. My
> memory holds the past and yet paradoxically
> knows, at the same time, that the past is gone and
> will never come again. It cherishes dead faces and
> silenced voices, yes, and lost evenings of child-
> hood. In some odd nonspatial way it contains
> houses and rooms that have been torn timber from
> timber and brick from brick. These have a greater
> permanence in that midge dance which contains
> them than ever they had in the world of reality.

III

Planning is rooted in our intuitive confidence in the infinite: we believe that we can remember forever. This is a peculiarly human hope and faith, and it allows us to do those special things that we consider sentient and especially human and creative, like mathematical logic.* Moreover, our confidence in infinity makes memories into truths—we hold fast our memories because they are the final certainties of the future. Robert Louis Stevenson wrote in *To any reader:*

---

*The French mathematician and physicist Henri Poincare [pp. 11-13] said:

> [One] cannot conceive [the] general truths [of Arithmetic] by direct intuition alone; to prove even the smallest theorem he must use reasoning by recurrence, for that is the only instrument which enables us to pass from the finite to the infinite...When we take in hand the general theorem, [then reasoning by recurrence] becomes indispensable...and without [our confidence in infinite repetitions] there would be no science at all, because there would be nothing general...
>
> The rule of reasoning by recurrence is irreducible to [other logical principles such as] the principle of contradiction. Nor can the rule come to us from experiment... When it is a question of a single formula to embrace an infinite number of syllogisms...experiment is powerless to aid. This rule, inaccessible to analytical proof and to experiment, is the [archetypic] a priori synthetic intuition.

As from the house your mother sees
You playing round the garden trees,
So you may see, if you will look
Through the windows of this book,
Another child, far, far away,
And in another garden, play.
But do not think you can at all,
By knocking on the window, call
That child to hear you. He intent
Is all on his play-business bent.
He does not hear; he will not look,
Nor yet be lured out of his book.
For, long ago, the truth to say,
He has grown up and gone away,
And it is but a child of air
That lingers in the garden there.

"He does not hear; he will not look, nor yet be lured out of his book." This little child is *ever* wild and free and filled with wonder, pottering about under rocks by hawkweed, bur-cucumber, chickory flowers, and oaktree seeds. Why, hours slip along without his noticing it—and each time we reread the poem, he lightly steps again alone but forever sure as coming home.

The child is remembered in the poem—and what is the difference between a poem and a memory? John Keats wrote that "poetry should surprise by a fine excess, and not by singularity; it should strike the reader as a wording of his own highest thought, and appear almost

---

Why then is this [rule] imposed upon us with such an irresistible weight of evidence? It is because it is only the affirmation of the power of the mind which knows it can conceive of the indefinite repetition of the same act, when the act is once possible. The mind has a direct intuition of this power, and experiment can only be for [the mind] an opportunity of using [this power], and thereby of becoming conscious of it.

a remembrance." My earliest remembrances are lightly hazy snapshots: they are whole wide scenes, bright pictures of rooms and rainy streets and huge dark houses.

I remember one day—
  It's in a springtime
That's rainlit and green-grey
  With cloud-covered skies.

I am standing outside
  A child all alone
Small on a sidewalk
  Near my home.

The trees' leaves are dark
  And round like green
Pebbles, and the bark
  Has a silver-grey sheen

That's finely etched
  With lines of brown.
The grass is wet,
  The walks stretch down

The streetside forever,
  And as far as I can see
The street has no ending.
  A large oak tree

Stands along the misty street
  Alive and watching me.
No one else is all around
  And thoughtful and remotely

Massive brown brick houses
  Peer from deep within
Their dark front porches;
  They loom in rows akin

To a giant stone orchard
  Of temples old and wise,
Solemn but not forbidding,
  Each mysterious and warm inside.

My memory is but a poem written somewhere in my mind. But is it accurate? Is it true, this memory of mine? I am certain that I have not recreated the scene exactly as others would know it, but I do not doubt that what I remember is close to what I saw with my infant eyes and what I felt in my little child soul—the continual faithful recurrence of the memory year after year gives me this trust. The exact details of a memory need not be perfectly reproduced for it to be true and certain. In the end, memory is only repetition: "Memory" comes from the old French *memorie* and *memorie* came to France from the Latin *memore* meaning to recount, to remember, to recall; "repetition" comes from the old French *repeter* meaning to fetch back, to seek the return of, to recall. At heart and from its own childhood, Memory is Repetition...

Memory is repetition; it is re-creation. How faithful must be this re-creation? For memory, there must be a certain "center" that recurs, but edges and shades may vary—we cannot help refinishing the boundaries, corners, and extensions of our images as we recreate them. Sometimes the trees in my memory have leaves like pebbles, sometimes the leaves are mistier. At times, the houses loom larger; some days, rain completely obscures the end of the street, while other days, the pavement disappears like a far mountain road in the distance. Memory is re-creation, and as we re-create we also create anew; memories are multiform and protean. But, if a memory is true, should it not be perfect and stereotyped? The notion that all true re-creations must be identical frozen crystalline images is a recent invention; it is a direct product of our writing culture,* and it has been reinforced by the high-fidelity reproductions of photography. Like a poem, memory is repetition, faithful re-creation. The difference between a poem—a written poem—and a memory is how precise the details must be in each re-creation: a crystalline determinacy is written poetry, a gentle determinancy is memory.

---

*Albert Lord, folklorist and historian [pp. 124 and 138]:

> The art of narrative song was perfected...long before the advent of writing. It has no need of stylus or brush to become a complete artistic and literary medium...But

IV

Planning is memory, and memory is repetitive re-creation. Faithful repetition is history, and it is memory; it is the past writ into the future, and it is concrete and operational. Repetition is planning, repetition is memory, repetition bridges past to future; and, for these reasons, repetition is a fundament of teaching and of learning. Iteration underlies teaching, because repetitive re-creation is an elemental form of understanding. We understand what we can repeat, and repetition lulls us into understanding. In regard to music, Anton Webern [p. 22] wrote:

> I said [in my last lecture] that the first principle is comprehensibility! How is it expressed [in this particular musical passage]? What strikes us first? The repetition! We find it almost childish. What's the easiest way to ensure comprehensibility? Repetition. All formal construction is built upon it, all musical forms are based on this principle.

Such understandings—understandings based on repetitive re-creation, faithful and determinate—also underlie all of science. Experimental results must be repeatable: the shape of a neutron's path, the position of a spot on a polyacrylimide gel, the dimensions of a femur, the number of bristles on the back of a fly—these phenomena

---

> writing, with all its mystery, came to the singer's people, and eventually someone approached the singer and asked him to tell the song so that he could write down the words...A written text was thus made of the words of [the] song [and] a fixed text was established...The change has been from stability of essential story, which is the goal of oral tradition, to stability of text, of the exact words of the story. The spread of the concept of fixity among the carriers of oral traditional epic is...one aspect of the transition from an oral society to a written society.

must recur if they are to find a place in the archives of science. Each scientific observation must be verifiable. Moreover, the promise of science is prediction—it is the promise of repetition. A scientific statement is useful only insofar as it holds true for the future. The scientific statement "F=ma" predicts: every time in the future when one assesses the force exerted by a moving object, he will find it equal to the mass of the object times its acceleration. The scientific statement "our genetic inheritance is encoded in the DNA of our germ cells" predicts: every time in the future when one examines the DNA of parents' germ cells, he will find that it specifies the child's genetic endowment.

Repetition is an all-pervasive necessity for science. "There is no science but the science of the general," wrote Poincare [p. 4]. "The object of the exact sciences is to dispense with...direct verifications [of particular cases]." These general statements—the fundaments of science—are inferred by induction, and induction is an operation requiring high-fidelity repetition. For instance, consider the classic use of induction, the general definition of Addition:

Addition operationally identifies the integer "x+n" in the statement "x + n = x+n." The inductive definition of Addition begins with a long line of integers in numerical order, a number line:

$$\ldots\ 0\ 1\ 2\ 3\ 4 \ldots n-1\ n\ n+1 \ldots$$

Each integer can be symbolized by x, and its rightmost neighbor is then symbolized by $x+1$. Now:

> *Positive Addition,* "x + n = x+n," is an operation that uses a given positive integer n to transform an arbitrary integer x into another larger integer $x+n$ farther along the number line.

The meaning of this general definition—our understanding of simple Addition for all particular cases—is built with induction; we understand Addition by repetition. First, we assume that we can carry out one particular axiomatic stepping operation:

$$x + 1 = x+1$$

which transforms any integer x into its rightmost neighbor $x+1$. Operationally, this moves us one step to the right on the number line.

Next, we start at some arbitrary place x along the number line, and we carry out a series of these stepping operations, repeating them a particular number *(n)* of times:

$$
\begin{array}{ll}
(0) & x = x \\
(1) & x + 1 = x+1 \\
(2) & x+1 + 1 = x+1+1 \\
(3) & x+1+1 + 1 = x+1+1+1 \\
 & \ldots \\
(n) & x+1+1+1+ \ldots + 1 + 1 = x+1+1+1+ \ldots +1+1
\end{array}
$$

Each left-hand side of the equation is simply the right-hand side of the preceding equation plus one stepping operation. Therefore, step (n) can be rewritten as n repetitions of our axiomatic operation $x + 1 = x + 1$:

$$(n) \qquad x + 1 + 1 + 1 + \ldots + 1 + 1 = x+1+1+1+\ldots+1+1$$

where both sides of the equation have n 1's. Step (n) is a summary of n repetitions of a single operation. In this way, the operational definition becomes:

Positive Addition "$x + n = x+n$" equates an operation "$x + n$" and an integer "$x+n$," defined as—

(1) "$x + n$" means faithfully repeating the single stepping operation "$x + 1$" n times along the number line.
(The full operation "$x + n$" is symbolized as: $x + n = x + 1 + 1 + 1 + \ldots + 1 + 1$ where there are n 1's.)

(2) "$x+n$" is the new integer position reached along the number line after n repetitions of the single stepping operation "$x + 1$."
(The integer "$x+n$" can be symbolized as: $x+n = x+1+1+1+\ldots+1+1$ where there are n 1's.)

In this definition, there are *n* repetitions, and there is no limit to the size of *n*. *N* can be one, one hundred, or one billion—it can be infinite. The definition is general. It holds for any *n* solely because of an inner human confidence: we believe that we will be able to repeat faithfully and forever the axiomatic operation "x + 1 = x+1." Faithful repetition is determinate re-creation; we trust in the determinacy of repetition, and upon this human conceit depends the power of mathematics.* Like mathematics, all of science ultimately proceeds on our confidence in stereotyped iteration. Scientists need the power of accurate repetition; full logic requires an endless future—an infinity—and scientific logic builds its precise predictions upon confidence in the future. Scientific logic is a methodical form of detailed planning; and, operationally, detailed planning is determinate re-creation and stereotyped repetition.

V

What builds this form of understanding, this understanding-by-secure-repetition? Where does determinate re-creation come from? What assures us that future patterns will be faithful *re*currences and not new occurrences? How can we trust in determinate repetition?

The creation of all patterns ranges along a spectrum, and pattern determinacy can be established by two general mechanisms. At one end of the pattern-assembly spectrum, determinate re-creation stems

---

*Each of the axiomatic operations of Addition—operations such as "x+ 1 = x+1"—is by itself

> A purely analytical instrument...incapable of teaching us anything new. If mathematics had no other instrument, it would immediately be arrested in its development; but it has recourse to [induction, i.e. to] reasoning by recurrence, [thus] it can continue its forward march. The essential characteristic of reasoning by recurrence is that it contains, condensed...in a single formula, an infinite number of syllogisms...[Reasoning by] induction is only possible if the same operation can be [faithfully] repeated indefinitely.
>
> [Poincare, p. 9]

from a ubiquitous mechanism, a mechanism that operates automatically. Farther out along the spectrum, an additional mechanism must be invoked specially to ensure determinate re-creation.

At the one end of the spectrum—at the simple end—some patterns can only be assembled into one particular configuration. A hand and a glove fit together in only one particular configuration, most jigsaw puzzles can be put together into only one final picture; similarly, take apart a complex mechanical clock, and you will find that there is only one way to assemble the pieces into a functional whole. In terms of configurational information, mechanical clocks, most jigsaw puzzles, and a hand and a glove are like ice crystals: freezing water under normal temperatures and pressures always produces ice crystals (ice I) of the same hexagonal shape. The inherent properties of the pattern-forming elements—the tetrahedral water molecules—determine completely the final pattern. Faithful repetition (determinacy in the re-creation of ice crystals) is dictated by the "universal assembly laws" that constrain the pattern-assembly system to construct only one configuration. Because the architecture of such patterns resides uniquely in the elements, these patterns are self-assembling.

In contrast to ice, solutions of silicon carbide (SiC) form a variety of crystalline lattices, ranging from cubic to hexagonal; even with exquisite control of temperature, pressure, and the microscopic texture of nucleation surfaces, one is hard pressed to fabricate pure silicon carbide crystals of any one particular form. Silicon carbide molecules are configurationally naive: their inherent properties permit a range of final crystal patterns and only the application of carefully engineered external templets can begin to constrain the crystal patterns of silicon carbide to any one predetermined type.

Silicon carbide crystals are at the far end of the spectrum of pattern-assembly systems, where patterns are built of elements that can be put together in every which way. Chess patterns, for instance—endless different configurations can arise during a chess match, and this distinguishes chess from games like checkers that can easily be programmed into a robot. Like silicon carbide crystals and chess configurations, many other patterns are inherently broad-minded and live at the far end—the complex end—of the pattern-assembly spectrum. Imagine a jigsaw puzzle in which all pieces are same-sized hexagons; here, the number of final patterns would be fairly staggering. Or consider an optical kaleidoscope or a box of multicolored marbles—shake each and a different design appears. In all these cases, constraints inherent in the unit elements are insufficient to limit the overall final patterns to any one particular configuration.

By dint of the inherent natures of their elements, patterns can be arrayed along a spectrum. At one end, the patterns are narrow-minded; at the other end, they are broad-minded—but in all cases, the patterns can nonetheless be made predictable. What, then, are the general sources of pattern predictability? One way to produce predictability, to ensure faithful recurrence, to build memories, and to enact planning is to create patterns (like a hand and a glove, typical jigsaw puzzles, mechanical clocks, and ice crystals) that are on the self-assembling end of the spectrum. These patterns form predictably, and their predictability stems from their single-minded elements. Single-minded self-assembly ensures repetition.

But what of the broad-minded patterns—those patterns whose elements come without narrow preconceptions about which particular configuration they will form? Sometimes, as in the case of juggling marbles in a box or cloud shapes in the sky or noises in the city or fingerprints or monkeys madly typing manuscripts, details of the final configurations are unpredictable; these patterns are random, recurrence is unfaithful, memory is feeble, and planning is absent. For random patterns, we cannot trust in stereotyped repetition. However, equally complex and inherently broad-minded patterns can also be predictable. Chess patterns, for instance, are complex and inherently broad-minded: if a child were to change the position of one chessman in its initial board position, he could produce as many as 1024 different configurations. But if a chess player makes the initial chess move, he will produce only one of 40 different configurations because the rules of play limit the actual first-move patterns. Rules, externally imposed guidelines—templets—constrain the configurationally naive pattern elements; and forever in our chess-bound universe, we can confidently expect to encounter only one twenty-fifth of the conceivable first-move chess patterns. External templets impose predictability on chess patterns: templets can ensure repetition in such inherently broad-minded patterns.

Like chess patterns, DNAs are also built from broad-minded, inherently unfettered elements. Nucleotides are the elemental units composing DNA patterns, and nucleotides can be strung together in any conceivable sequence; yet DNAs are the classic examples of complex, exquisitely predictable natural patterns. DNA is genetic memory. Specific, distinct DNA patterns code for the one hundred thousand specific, distinct, and reproducible proteins in each human; and these DNA patterns recur: they are repeated and remembered precisely from cell to cell in a person as he develops and from cell to cell in his children as they grow. Although they are composed of configurationally naive elements, organismal DNAs faithfully recur. Here faithful repetition is ensured by stable templets. New DNA molecules are assembled nucleotide by nucleotide along preexisting DNA molecules, and DNA patterns endure because new DNA molecules are fabricated via blueprints that last.

"When old age this generation shall waste, thou shalt remain." Repetition is endurance, and templets are the memory, the planning and the determinacy that forge predictability and endurance for broad-minded patterns. When pure, clean, homogeneous self-assembly cannot guarantee the inevitability of particular final patterns, persistent stable templets can nonetheless unerringly seal their fates.

VI

Prediction of future patterns can be made secure in two ways:

(a) The pattern elements can be so limited in their interactive potentials that only one configuration can ever form.

(b) Predictable repetitive patterns can be built using long-lasting templets.

No matter how configurationally knowledgeable or naive are the elements, the existence of a stable templet always provides a fine basis for secure prognostication. The world flows ever on and on, but is repetition in the cards? Will a particular pattern recur in the future? Will there be determinate re-creation? The answer is "yes" when the pattern's elements are finicky and single-minded, and it is "yes" when the pattern's templet is enduring and intransient.

People plan with both types of mechanisms—we ensure memorable patterns with self-assembling elements and also with stable templets. We remember the lengths of the twelve months because of a pattern of words that assemble snugly into one particular rhyming jingle; the numbers "self-assemble":

> Thirty days hath September
> April, June, and November;
> All the rest have thirty-one,
> Excepting February—when it's done
> There've been twenty-eight days cold and clear
> (And twenty-nine days in each leap year).

But we remember the first twenty-one digits of the number pi—3.14159265358979323846—via stable templets. To recreate pi, we do what I have just done; we go to the enduring templet on our bookshelf and look up the number in a table of mathematical constants.

To plan and to remember are to recreate, and it is our impulsion to recreate that drives the artist to plan and to remember. With what clay does he plan, of what does he build his most human of artworks, his artistic memories? Not the persnickety, lockblock elements that of their own single-minded organization will self-assemble only into one particular pattern. No, we build art with configurationally-naive atoms. When we create special human artworks in literature, dance, and music, in painting, sculpture, and architecture, we work with bits and pieces that for all they know can associate and intermingle in any number of strange, wondrous, and varied assemblages. The notes that open the "Hallelujah Chorus" in George Fredrich Handel's oratorio *The Messiah,*

and the words that begin e.e. cummings's poem "what if a much of a which of a wind":

> what if a much of a which of a wind
> gives the truth to summer's lie;
> bloodies with dizzying leaves the sun
> and yanks immortal stars awry?
> Blow king to beggar and queen to seem
> (blow friend to fiend: blow space to time)
> —when skies are hanged and oceans drowned,
> the single secret will still be man

—these elemental units are endlessly configurable.

In artistic patterns, the elements are innocent children; wide-eyed and trusting, they will queue in any sequence. Artistic units can be arranged in almost any pattern, and the greatness of Bach's music and of Auden's poetry lies not in the elements themselves. Words and body movements and musical notes are infinitely configurable elements like multicolored marbles: when shaken together, they fall into all imaginable patterns. Therefore, artistic creations must be extensively templeted. Artistic majesty resides in the particular configurations that are externally imposed by the creators, and the human artistry of creation comes from its human templets.

Art is extensively templeted—it is also memorable. Memorable means determinate and predictable, and predictability is one feature that distinguishes our art from ephemeral happenstance. The random typing of a monkey may well produce a sonnet, but the monkey's sonnet will disappear amidst the mountain of nonsense he types alongside. Artistic sonnets, on the other hand, are filtered from the chaff, they are highlighted with a golden glow, they stand alone, they are memorable. With true artworks, we can get our hands on them time and time again; and, fundamentally, this means that they are determinate. Artistic sonnets can be faithfully recreated: art requires repetition. Because artistic creations are built of broad-minded, configurationally naive elements, art also requires stable templets; for an artistic creation to be memorable, for it to be predictable, recurrent, secure, and trustworthy, its templets must endure. In an oral culture, the song must be faithfully passed from mind to mind; in a writing culture, the poem must be inscribed in a book. Secure, faithful repetition is the fount of understanding, and, in the human arts, understanding-by-secure-repetition builds on the rock of invariant templets.

VII

> O joy! that in our embers
> Is something that doth live,
> That nature yet remembers
> What was so fugitive!
>
> [Wordsworth]

Like her human children, Nature plans, and she uses both types of planning mechanisms: she ensures memorable patterns with self-assembling elements and also with stable templets. Ice crystals are determinate re-creations because the water molecules self-assemble into only one particular configuration; in contrast, hemoglobin molecules are determinate re-creations because they are built from long-lasting DNA templets. Can Nature abstract or theorize? I am not sure, but Nature can certainly plan. If there is a natural teleology—and I think that there is—then it flows from the secure prognostication that is built on faithful recurrence, stereotyped re-creation, and trustworthy repetition.

Nature plans, of that I am certain; natural teleology cannot be lightly dismissed. An early spring rough lawn, can Nature remember back? It was a summer evening—three children alone at night, and we were playing "Mother-may-I." Nature was Mother, the Queen on our chessboard lawn; we got to our places and called out: "How many? What kind? How many steps do we find?" And I was just a tiny child, so she told me to take two baby steps and then one little baby umbrella step for the joy of a summer evening. "Mother-may-I?" Yes, you may. Mother Nature stood at one edge of the world, and we began at the other; slowly and swiftly, saltarello terpsichore through the twilight we bounced and danced, until one of us reached out and touched her. It was chess, and in the eighth rank a pawn became a queen. Was it all determined? Were the steps fated? Was the way measured, was the end one? Had planning transformed the indeterminate haze of tomorrow into a determinate destiny of today?

"Mother-may-I?" I called out. Did the answer echo back?

Did Nature plan for me? Could she prognosticate with the arcane sentience that we pride ourselves is only human? Did she contrive with a willful creation that we would call "artistic"? Nature plans ice crystals, but it does not feel mysterious that Nature sometimes ensures the future through such self-assembling systems. The realm of physics offers many examples: besides ice crystals, there are standing waves oscillating with natural harmonic frequencies, hexagonally-packed soap bubbles, and spherical surfaces. To a large extent, these recurrent patterns can be explained on the basis of inherently single-minded unit elements—elements constrained to particular global configurations by the "universal assembly laws" of physics. This is faithful repetition through self-assembly, and it does not threaten the uniquely human artistry of creation. Order need not be imposed from the outside, sentience need never be invoked, and artistry is not at issue. It is more unsettling, though, when Nature recreates complex patterns via extensive templets. Repetition by templeting means that the form information—the order and the configuration of the particular final pattern—is not inherent in the elements that make up the pattern: form, shape, organization, and lineaments are imposed from the outside. For Nature, this type of creation is nowhere more dramatic than in the realm of biology. Life is the stereotyped re-creation of complex patterns, and life comes from extensive templets. Nature's organisms are maximally-complex determinate patterns: most organisms require maximally determinate templets for their assembly.

Like humans, Nature recreates with extensive templets, and some of Nature's templets are stable, enduring, and predictable. Predictability is one feature that distinguishes our art from random happenstance. What distinguishes our art from Nature's extensively templeted recurrent patterns? Art is something that lasts. But rocks last and meteors persist and the universe endures—longevity alone is not the essential criterion. The longevity of art is an indirect characteristic, and it reflects the stability and the endurance of the artistic templets: artworks last because their templets persist. This is the *mode* of art's longevity, but it is not the *cause* of its artistry; longevity is a necessary but not a sufficient characteristic of art. Only *certain* extensively templeted recurrent creations are true artworks. A bad poem can be as stably templeted as a good poem. Both can be written side-by-side in the same book, well-bound and printed on acid-free paper. Both poems will last physically, but only one poem—the poem that is good in human terms, the poem that touches our human soul—will endure artistically. A good poem has artistic perpetuity because we will continue to read it and to repeat it. Through the years and generation after generation, physically persistent poems are tested, and the poem that resonates most soundly within our soul, that nestles most fitly into our deep emotional homes, and that, as John Keats wrote, appears "almost a remembrance"—that poem *becomes* an artistic remembrance. It is humanly memorable, it is humanly repeatable, it is human art. As Odysseus said: "Even one's griefs touch our soul and become a joy when remembered in our heart again and again in old age from all of the deepest things that one has wrought and endured."

In the land of the living, Nature sometimes plans through enduring extensive templets, and this is disquieting because extensive templets also characterize artistic creation. Human artworks are extensively templeted: they are patterns fabricated of infinitely configurable units, like words and body movements and musical notes. To sculpt particular creations from such endlessly conforming media, we must use full templets—we impose the entire final form from the outside. It is human whim and will that completely templet a poem, a dance, and a sonata, but whose whim and will have templeted a child?

Artistic creations are extensively templeted, and memorable art is fabricated from *stable* templets. The organizational frames, the detailed patterning information, the particular formats for the words and the body movements and the notes of an artwork—these endure. They can be passed from hand to hand, and they can be bequeathed from mother to child. Any re-creation that is memorable has this longevity called "determinacy," and artistic creation is no exception. Artistic creations are determinate—they must be predictable, for predictability is one way that art differs from random happenstance.

Like the human artist, Nature, too, generates extensively templeted re-creations, and some of these re-creations are enduring, repeatable, and complex patterns. What distinguishes the lush recurrent natural skeins and tangles from human art? Nature's patterns are as complex and as repeatable as human art; therefore, the distinction between Nature's complex recurrent patterns and human art is neither complexity nor longevity. This distinction can only be the extent to which these recurrent templeted patterns touch our human souls. But Nature's most extensively templeted recreations are our children, playing on a little spring rough lawn early one summer evening; little children ever wild and free and filled with wonder, pottering about under rocks by hawkweed, bur-cucumber, chickory flowers, and oaktree seeds—why, hours slip along without their noticing it, and each summer evening a child lightly steps again alone but forever sure as coming home. And, if a child does not touch our souls, then what does?

BIBLIOGRAPHY

Introduction

Adam, G. and M. Delbruck. "Reduction of Dimensionality of Biological Diffusion Processes". Pp. 198-215 in *Structural Chemistry and Molecular Biology.* eds. A. Rich and N. Davidson. San Francisco: W.H Freedman. 1968.

Ash, M.M., Jr. *Wheeler's Dental Anatomy, Physiology, and Occlusion.* Philadelphia: W.B. Saunders. 1984.

Bohm, D. *Causality and Chance in Modern Physics.* Philadelphia: University Press. 1971.

Bronowski, J. "Science as Foresight." Pp. 385-436 in *What is Science?* ed. J.R. Newman. New York: Simon and Schuster. 1955.

Eiseley, L. "The Secret of Life." *The Immense Journey.* New York: Vintage Books. 1959.

Grenander, U. *Pattern Synthesis.* New York: Springer-Verlag. 1976

Grenander, U. *Pattern Analysis.* New York: Springer-Verlag. 1978.

Grenander, U. *Regular Structures.* New York: Springer-Verlag. 1981.

Harary, F. *Graph Theory.* Reading, Mass.: Addison-Wesley. 1969.

Huxley, A.F. "Looking Back on Muscle." Pp. 23-64 in *The Pursuit of Nature.* eds. A.L Hodgkin, A.F. Huxley, W. Feldberg, W.A.H. Rushton, R.A. Gregory, and R.A. McCance. Cambridge: Cambridge University Press. 1977.

Katz, M.J. 1986. *Templets and the Explanation of Complex Patterns.* Cambridge: Cambridge University Press. 1986.

Katz, M.J. and Y.-S. Chow. *Journal of Theoretical Biology* 113: 1-13. 1985.

Keele, K.D. *Leonardo da Vinci's Elements of the Science of Man.* New York: Academic. 1983.

Kellogg, R. *Analyzing Childen's Art.* Mayfield Publishing Company. 1970.

Kraus, B.S., R.E. Jordan, and L. Abrams. "A Study of the Masticatory System." *Dental Anatomy and Occlusion.* Baltimore: Williams & Wilkins. 1969.

Laplace, P.S. de 1814. *A Philosophical Essay on Probabilities*. trans. W. Truscott and F.L. Emory. New York: Dover Publications. 1951.

Mandelbrot, B.B. *The Fractal Geometry of Nature*. New York: W.H. Freeman. 1983.

Monod, J. *Chance and Necessity*. trans. A. Wainhouse. New York: Vintage Books. Pp. 96-98. 1971

Poincare, H. *Science and Hypothesis*. New York: Dover Publications. 1952.

Polya, G. *How to Solve It: A New Aspect on Mathematical Method*. 2nd edition. Garden City, N.Y.: Doubleday Anchor. 1957.

Reichenbach, H. "The Philosophical Significance of the Theory of Relativity." Pp. 289-311 in *Albert Einstein: Philosopher-Scientist*. ed. PA. Schlipp. La Salle, Ill.: Open Court. 1949.

Shafer, W.G., M.K. Hine, and B.M. Levy. *A Textbook of Oral Pathology*. 4th edition. Philadelphia: W.B. Saunders. 1983.

Stevens, P.S. *Handbook of Regular Patterns: An Introduction to Symmetry in Two Dimensions*. Cambridge, Mass.: M.I.T. Press. 1980.

Thompson, D.W. *Growth and Form*. 2nd edition. New York: The Macmillan Co. 1942.

PATTERN CONCEPTS

Chapter 1

Abbott, E.A. 1884. *Flatland: A Romance of Many Dimensions*. New York: Barnes & Noble. 1983.

Anson, B.J. *An Atlas of Human Anatomy*. 2nd edition. Philadelphia: W.B. Saunders. 1963

Cameron, E. *The Wonderful Flight to the Mushroom Planet*. Boston: Little, Brown & Co. 1954.

Culoti, J.G., G. von Ehrenstein, M.R. Culotti, and R.L. Russell. "A Second Class of Acetylcholinesterase-Deficient Mutants of the Nematode *Caenorhabditis elegans.*" *Genetics* 97: 281-305. 1981.

Cummins, H. and C. Midlo. *Finger Prints, Palms and Soles.* Philadelphia: Blakiston. 1943.

Danford, D.E. and H.W. Munro. "Water-Soluble Vitamins: The Vitamin B Complex and Ascorbic Acid." Pp. 1576-1582 in *The Pharmacological Basis of Therapeutics.* 6th Edition. eds. A.G. Gilman, L.S. Goodman, and A. Gilman. New York: The Macmillan Co. 1980.

Doyle, A. Conan. *The Sign of the Four.* New York: Penguin Books. 1982.

Dumont, J.P.C. and R.M. Robertson. "Neuronal Circuits: An Evolutionary Perspective." *Science* 233: 849-853. 1986.

Eiseley, L. "The Slit." *The Immense Journey.* New York: Vintage Books. 1959.

Hawkey, C. "Hemostasis in Mammals." Pp. 69-86 in *Workshop on Animal Models of Thrombosis and Hemorrhagic Diseases.* DHEW publ. no. NIH 76-982. Washington, D.C.: The National Academy of Sciences (USA). 1975.

Jacob, F. "Evolution and Tinkering." *Science* 196: 1161-1166. 1977.

Katz, M.J. *Templets and the Explanation of Complex Patterns.* Cambridge: Cambridge University Press. 1986.

Katz, M.J. and U. Grenander. "Developmental Matching and the Numerical Matching Hypothesis for Neuronal Cell Death." *Journal of Theoretical Biology* 98: 501-517. 1982.

Katz, M.J. and R.J. Lasek. "Evolution of the Nervous System: Role of Ontogenetic Mechanisms in the Evolution of Matching Populations." *Proceedings of the National Academy of Sciences (USA)* 75 1349-1352. 1978.

Katz, M.J., R.J. Lasek, and I.R. Kaiserman-Abramof. "Ontophyletics of the Nervous System: Eyeless Mutants Illustrate How Ontogenetic Buffer Mechanisms Channel Evolution." *Proceedings of the National Academy of Sciences (USA)* 78: 397-401. 1981.

Katz, M.J., R.J. Lasek, and J. Silver. "Ontophyletics of the Nervous System: Development of the Corpus Collosum and Evolution of Axon Tracts." *Proceedings of the National Academy of Sciences (USA)* 80: 5936-5940. 1983.

Lehninger, A.L. *Biochemistry.* 2nd edition. New York: Worth. Pp. 631-632. 1975.

Linzen, B., N.M. Soeter, A.F. Riggs, H.-J. Schneider, W. Schartau, M.D. Moore, E. Yokota, P.Q. Behrens, H. Nakashima, T. Takagi, T. Nemoto, J.M. Vereijken, H.J. Bak, J.J. Beintema, A. Volbeda, W.P.J. Gaykema, and W.G.J. Hol. "The Structure of Arthropod Hemocyanins." *Science* 229: 519-524. 1985.

Lucretius. *The Nature of the Universe.* trans. R.E. Latham, Book 4. Lines 823-876. Baltimore: Penguin Classics. 1951.

Needham, A.E. *The Uniqueness of Biological Materials.* Oxford: Pergamon Press. 1965.

Ratnoff, O.D. *Bleeding Syndromes: A Clinical Manual.* Springfield, Illinois: Charles C. Thomas. 1960.

Shapley, R. and J. Victor. "Hyperacuity in Cat Retinal Ganglion Cells." *Science* 231: 999-1002. 1986.

Sperry, R. "Some Effects of Disconnecting the Cerebral Hemispheres." *Science* 217: 1223-1226. 1982.

Thorn, R.G. and G.L. Barron. "Carnivorous Mushrooms." *Science* 224: 76-78.

Walsby, A.E. "A Square Bacterium." *Nature* 283:69-71. 1980.

Williams, R.J. *Biochemical Individuality.* New York: John Wiley & Sons. 1956.

Chapter 2

Adelmann, H.B. *The Embryological Treatises of Hieronymus Fabricius of Aquapendente.* Ithaca, N.Y.: Cornell University Press. 1967.

Chaitin, G.J. "Randomness and Mathematical Proof." *Scientific American* 232: 47-52. 1975.

Chaitin, G.J. "Algorithmic Information Theory." Pp. 38-41 in *Encyclopedia of Statistical Sciences, Vol. 1.* eds. S. Kotz and N.L. Johnson. New York: John Wiley & Sons. 1982.

"complex" (entry). *Oxford English Dictionary.* Oxford: Oxford University Press, 1971.

Karmel, B.Z. and E.B. Maisel. "A Neuronal Activity Model for Infant Visual Attention." Pp. 77-131 (esp. p. 79) in *Infant Perception: From Sensation to Cognition, Volume 1.* eds. L.B. Cohen and P. Salapatek. New York: Academic Press. 1975.

Katz, M.J. *Templets and the Explanation of Complex Patterns.* Cambridge: Cambridge University Press. 1986.

Martin-Lof, P. "The Definition of Random Sequences." *Information and Control* 9: 602-619. 1966.

Popper, K.R. *The Logic of Scientific Discovery.* New York: Basic Books. Pp. 136-145. 1959.

Sandburg, C. "How to tell corn fairies if you see 'em." *Rootabaga Stories, Part One.* San Diego: Harcourt Brace Jovanovich, 1950. Pp. 205-212. 1922.

"simple" (entry). *Oxford English Dictionary.* Oxford: Oxford University Press. 1971.

Williams, R.J. *You Are Extraordinary.* New York: Random House. 1967.

Chapter 3

Altman, P.L. and D.S. Dittmer, eds. *Biology Data Book,* 2nd edition. Volumes 1-3. Bethesda, Maryland: Federation of the American Society for Experimental Biology. 1972-74

Anson, B.J. *An Atlas of Human Anatomy.* 2nd edition. Philadelphia: W.B. Saunders. 1963.

Aristotle. "On the Soul" Book 3, Chapter 9, Number 21. Pp.145-235 in *Introduction to Aristotle.* ed. R. McKeon. New York: Modern Library. 1947.

Bacon, F. "Novum Organum or True Suggestions for the Interpretation of Nature" (1620). Pp. 454-540 in *Selected Writings of Francis Bacon,* ed. H.G. Dick. New York: Modern Library. 1955.

Bonnet, C. "La Palingenesi Philosophique" (1769). Translated and quoted on p. 25 in S.J. Gould, *Ontogeny and Phylogeny.* Cambridge, Mass.: Harvard University Press. 1977.

Bridgman, P.W. *The Nature of Physical Theory.* Princeton: Princeton University Press. 1936.

Brock, T.D. "Life at High Temperatures." *Science* 230: 132-138.

Chaitin, G.J. "On the Length of Programs for Computing Finite Binary Sequences." *Journal fo the Association for Computing Machines* 13: 547-569. 1966.

Chaitin, G.J. "Algorithmic Information Theory." *I.B.M. Journal of Research and Development* 21: 350-359. 1977.

cummings, e.e. "neither could say." In *e.e. cummings. Poems 1923-1954.* New York: Harcourt, Brace & World.

Darwin, C. "Recapitulation and Conclusion" (1872). *The Origin of Species.* 6th Edition. New York: Modern Library.

Einstein, A.E. "On the Electrodynamics of Moving Bodies" (1923). In *The Principle of Relativity.* eds. H.A. Lorentz, A. Einstein, H. Minkowski, and H. Weyl. New York: Dover Publications.

Einstein, A.E. "On the Method of Theoretical Physics" (1934). *Philosophy of Science* 1: 163-169.

Galen. *On the Natural Faculties.* Book 1, Chapter 6. trans. A.J. Brock. London: William Heinemann. 1916.

Gleizes, A. and J. Metzinger (1912). Pp. 1-18 in *Modern Artists on Art.* ed. R.L. Herbert. Englewood Cliffs, N.J.: Prentice-Hall. 1964.

Haeckel, E. *Anthropogenie oder Entwickelungsgeschichte des Menschen.* Leipzig: Verlag von Wilhelm Engelmann. 1910.

"Ernst Heinrich Haeckel" (entry). *The Encyclopedia Britannica.* 11th Edition. Volume 12. Pp. 803-804. 1911.

Jerne, N.K. "The Generative Grammer of the Immune System." *Science* 229: 1057-1059. 1985.

Johnson, J.W. "The Creation." *God's Trombones: Seven Negro Sermons in Verse.* New York: Penguin Books. Pp. 17-20. 1969.

Kant, I. *Critique of Pure Reason* (1781). trans. F.M. Muller. Garden City, N.Y.: Anchor Books. P. 415.

Klee, P. *On Modern Art.* London: Faber & Faber. 1948.

Kolmogorov, A.N. "Three Approaches to the Quantitiative Definition of Information." *Problems in Information Transmission* 1: 1-17. 1965.

Kolmogorov, A.N. "Three Approaches to the Quantitative Defini-Probability Theory. " *I.E.E.E. Transactions in Information Theory.* IT-14: 662-664. 1968.

Leibniz, G.W. "The Principles of Nature and Grace, Based on Reason" (1714). Pp. 528-529 in *Leibniz: Selections.* New York: Charles Scribner's Sons. 1951.

Locke, J. 1676. *Essays on the Law of Nature* (1676). P. 109. London: Oxford University Press. 1954.

Martin-Lof, P. "The Definition of Random Sequences." *Information and Control* 9: 602-619. 1966.

McWhirter, N., ed. *Guinness Book of World Records.* New York: Bantam Books. 1985.

Monod, J. *Chance and Necessity.* trans. A. Wainhouse. New York: Vintage Books. 1971.

Muzzio, N. and J.R. Newman. *Godel's Proof.* New York: New York University Press. 1958.

Newton, I. 1729. "The Rules of Reasoning in Philosophy" (1729). Book III, Rule I. *The Mathematical Principles of Natural Philosophy.* trans. A Motte. London: Dawsons of Pall Mall. 1968.

Perler, F., A. Efstratiadis, P. Lomedico, W. Gilbert, R. Kolodner, and J. Dodgson. "The Evolution of Genes: The Chicken Preproinsulin Gene." *Cell* 20: 555-566.

Piaget, J. *The Origin of Intelligence in Children,* trans. M. Cook. New York: International Universities Press. 1952.

Piaget, J. *The Child's Conception of Number.* New York: W.W. Norton. 1965.

Piaget, J. *The Child and Reality.* trans. A. Rosin. New York: Basic Books. 1973.

Piaget, J. and B. Inhelder. *The Psychology of the Child.* New York: Basic Books. 1969.

Plato. "Timaeus." trans. B. Jowett. Pp. 1151-1211 in *The Collected Dialogues of Plato, including the Letters.* eds. E. Hamilton and H. Cairns. Princeton: Princeton University Press. 1961.

Webern, A. *The Path to the New Music.* Bryn Mawr, Pa.: Theodore Presser. 1963.

Wigner, E. "The Unreasonable Effectiveness of Mathematics in the Natural Sciences." Pp. 222-237 in *Symmetries and Reflections: Scientific Essays of Eugene P. Wigner.* Bloomington: Indiana University Press. 1967.

Williams, R.J. *Biochemical Individuality.* New York: John Wiley & Sons. 1956.

Williams, R.J. *You Are Extraordinary.* New York: Random House. 1967.

Wilson, E.O. 1985. "Time to Revive Systematics." *Science* 230: 1227. 1985.

Chapter 4

Bell, E.T. *The Development of Mathematics.* 2nd Edition. New York: McGraw-Hill. P. 593. 1945.

Boole, G. *Collected Logical Works, Volume 2: The Laws of Thought* (1854). Chicago: Open Court Publishing Co. 1940.

Clerke, A.M. "Pierre Simon Laplace." *The Encyclopedia Britannica.* 11th edition. Volume 16, pp. 200-203. 1911.

Cowper, W. "The Task" Book II ("The Time-Piece," 1784). In *Cowper: Verse and Letters,* ed. B. Spiller. Cambridge, Mass.: Harvard University Press, 1968.

David, F.N. "Some Notes on Laplace." Pp. 30-44 in *Bernoulli, 1713; Bayes, 1763; Laplace, 1813. Anniversary Volume.* eds. L. Le Cam and J. Neyman. New York: Springer-Verlag. 1965.

Dickinson, E. "To Make a Prairie." P. 432 in *The Oxford Book of American Verse.* ed. F.O. Matthiessen. New York: Oxford University Press. 1950.

Feller, W. *An Introduction to Probability Theory and Its Applications.* 2nd Edition. New York: John Wiley & Sons. 1966.

Fisher, R.A. *The Design of Experiments.* 8th Edition. New York: Hafner Publishing Co. 1966.

Fitzgerald, R. (trans.) *Homer: The Odyssey.* Garden City, N.Y.: Anchor Books. 1963.

Freud, A. *Normality and Pathology in Childhood: Assessments of Development.* New York: International Universities Press. 1965.

Gnedenko, B.V. "The Concept of Probability." Pp. 21-87 in *The Theory of Probability,* Chapter I. trans B.D. Seckler. New York: Chelsea Publishing Co. 1968.

Katz, M.J. *Templets and the Explanation of Complex Patterns.* Cambridge: Cambridge University Press. 1986.

Katz, M.J. "Is Evolution Random?" In *Development as an Evolutionary Process: M.B.L. Lectures in Biology Series.* eds. R.A. Raff and E.C. Raff. New York: A.R. Liss. pp. 285-315. 1987.

Laplace, P.S. de. *A Philosophical Essay on Probabilities* (1814). trans. F.W. Truscott and F.L. Emory. New York: Dover Publications. 1951.

Le Cam, L. and J. Neyman. "Forward." Pp. iii-ix in *Bernoulli, 1713; Bayes, 1763; Laplace, 1813. Anniversary Volume.* eds. L. Le Cam and J. Neyman. New York: Springer-Verlag. 1965.

Kolmogorov, A.N. *Foundations of the Theory of Probability.* 2nd Edition. trans. N. Morrison. New York: Chelsea Publishing Co. 1956

Koopman, B.O. "The Bases of Probability." *Bulletin of the American Mathematical Society* 46: 763-774. 1940.

Martin-Lof, P. "The Definition of Random Sequences." *Information and Control* 9: 602-619. 1966.

Mitchell, J. "An Inquiry into the Probable Parallax, and Magnitude of the Fixed Stars from the Quantity of Light which They Afford Us, and the Particular Circumstances of their Situation" (1767). *Philosophical Transactions of the Royal Society (Lond.)* 57: 234-264.

Mill, J.S. *System of Logic, Ratiocinative and Inductive; Being a Connected View of the Principles of Evidence and the Methods of Scientific Investigation* (1846). Book 3, Chapters 17 and 18. New York: Harper & Brothers.

Ore, O. "Pascal and the Invention of Probability Theory." *American Mathematical Monthly* 67: 409-419. 1960.

"probability" (entry). *Encyclopedia Britannica.* 9th edition. London. Volume 19. 1885.

"random" (entry). *Oxford English Dictionary.* Oxford: Oxford University Press. 1971.

Scott, W. "The Lord of the Isles," Canto 5, number 18. Pp. 411-473 in *The Poetical Works of Sir Walter Scott.* ed. J.L. Robertson. London: Oxford University Press, 1964.

Stoppard, T. *Rosencrantz & Guildenstern Are Dead.* New York: Grove Press. 1967.

Tippett, L.H.C. *Tracts for Computers, No. 15: Random Sampling Numbers.* London: Cambridge University Press. 1927.

von Misis, R. *Probability, Statistics and Truth.* New York: The Macmillan Co. 1957.

Chapter 5

Bacon, F. "Novum Organum, or True Suggestions for the Interpretation of Nature" (1620). Pp. 454-540 in *Selected Writings of Francis Bacon.* ed. H.G. Dick. New York: Modern Library. 1955.

Banting, F.G. and C.H. Best. "Pancreatic Extracts." *Journal of Laboratory and Clinical Medicine* 7: 464-472. 1922.

Bridgman, P.W. "The Logic of Modern Physics" (1928). Pp. 34-46 in *Readings in the Philosophy of Science.* eds. H. Feigl and M. Brodbeck. New York: Appleton-Century-Crofts, 1953.

Bridgman, P.W. *The Nature of Physical Theory.* Princeton: Princeton University Press. 1936.

Bronowski, J. "Science as Foresight" (1955). Pp. 385-436 in *What Is Science?* ed. J.R. Newman. New York: Simon and Schuster.

Conant, J.B. *Science and Common Sense.* New Haven: Yale University Press. 1951.

Dodd, J.S., ed. *The ACS Style Manual.* Washington, D.C.: American Chemical Society. 1986.

Dorland, W.A.N. *The American Illustrated Medical Dictionary.* Philadelphia: Saunders. 1944.

Eddington, A.S. *The Nature of the Physical World.* New York: The Macmillan Co. 1929.

Einstein, A. "Autobiographical Notes" (1949). Pp. 3-95 in *Albert Einstein: Philosopher-Scientist.* ed. P.A. Schlipp. La Salle, Ill: Open Court Publishing Co. 1949.

Fowler, H.W. *A Dictionary of Modern English Usage.* London: Oxford University Press. 1927.

Fritzsch, H. *Quarks: The Stuff of Matter.* New York: Basic Books. 1983.

Giancoli, D.D. *General Physics.* Englewood Cliffs, N.J.: Prentice-Hall. 1984.

Hodgkin, A.L. "Chance and Design in Electrophysiology: An Informal Account of Certain Experiments on Nerves carried out between 1934 and 1952." Pp. 1-21 in *The*

Pursuit of Nature. eds. A.L. Hodgkin, A.F. Huxley, W. Feldberg, W.A.H. Rushton, R.A. Gregory, and R.A. McCance. Cambridge: Cambridge University Press. 1977.

Katz, M.J. *Templets and the Explanation of Complex Patterns.* Cambridge: Cambridge University Press. 1986.

Katz, M.J. *Elements of the Scientific Paper.* New Haven: Yale University Press. 1985.

Kuhn, T.S. *The Structure of Scientific Revolutions.* 2nd edition. Chicago: University of Chicago Press. 1970.

MacLeish, A. "Ars poetica." *Poems, 1924-1933.* Boston: Houghton Mifflin. 1933.

Morrison, R.T. and R.W. Boyd. *Organic Chemistry.* 4th edition. Boston: Allyn and Bacon. 1983.

O'Connor, M. and F.P. Woodford. *Writing Scientific Papers In English.* Amsterdam. 1978.

Oppenhiemer, R. "Communication and Comprehension of Scientific Knowledge" (1963). Pp. 271-279 in *The Scientific Endeavor.* New York: Rockefeller University Press.

Poincare, H. *Science and Hypothesis* (1903). New York: Dover Publications, 1952.

"science" (entry). *The Merriam-Webster Dictionary.* New York: Pocket Books, 1974.

Windholz, M., S. Budavari, R.F. Blumetti, and E.S. Otterbein, eds. 1983. *The Merck Index.* 10th edition. Rahway, N.J.: Merck & Co. P. 4866.

Chapter 6

a. *General*

Abderhalden, E. *Textbook of Physiological Chemistry in Thirty Lectures.* trans. W.T. Hall and G. Defren. New York: John Wiley & Sons. 1908.

Abel, J.J. *Chemistry in Relation to Biology & Medicine with Especial Reference to Insulin & Other Hormones.* Baltimore: Williams & Wilkins Co. 1939.

Aristotle. *On The Soul. Parva Naturalia. On Breath.* trans. W.S. Hett. Cambridge, Mass.: Harvard University Press. 1935.

Aristotle. *Parts of Animals.* Book 1, Chapter 5, lines 81-87. trans.

A.L. Peck. Cambridge, Mass.: Harvard University Press. 1937.

Bloch, K. "On the Evolution of a Biosynthetic Pathway." Pp. 143-150 in *Reflections on Biochemistry.* eds. A. Kornberg, L. Cornudella, B.L. Horecker, and J. Oro. Oxford: Pergamon Press. 1976.

Dayhoff, M.O., ed. *Atlas of Protein Sequence and Structure.* Volumes 1-5, Supplements 1-3. Silver Spring, Maryland: National Biomedical Research Foundation. 1965-1979.

Downes, H.R. *The Chemistry of Living Cells.* New York: Harper & Bros. 1955.

Galen. *On the Usefulness of the Parts of the Body* trans. M.T. May. Ithaca, N.Y.: Cornell University Press. 1968.

Jensen, H. and E.A. Evans, Jr. "The Chemistry of Insulin." *Physiological Reviews* 14: 188-209. 1934.

Katz, M.J. *Templets and the Explanation of Complex Patterns.* Cambridge: Cambridge University Press. 1986.

Keilin, D. *The History of Cell Respiration and Cytochrome.* Cambridge: Cambridge University Press. 1966.

Kornberg, A., L. Cornudella, B.L. Horecker, and J. Oro, eds. *Reflections on Biochemistry.* Oxford: Pergamon Press. 1976.

Leicester, H.M. *Development of Biochemical Concepts from Ancient to Modern Times.* Cambridge, Mass.: Harvard University Press.

Lehninger, A.L. *Biochemistry.* 2nd edition. New York: Worth. 1975.

Lloyd, D.J. "Recent Developments in our Knowledge of the Protein Molecule." Pp. 23-35 in *Perspectives in Biochemistry: Thirty-one Essays Presented to Sir Frederick Gowland Hopkins.* eds. J. Needham and D.E. Green. Cambridge: Cambridge University Press. 1937.

Mendelsohn, E. *Heat and Life: The Development of the Theory of Animal Heat.* Cambridge, Mass.: Harvard University Press. 1964.

Monod, J. *Chance and Necessity.* trans. A. Wainhouse. New York: Vintage Books. 1971.

Partington, J.R. *A History of Chemistry.* Volumes 1-4. London: MacMillan & Co. 1961-1970.

Plato. "Timaeus." trans. B. Jowett. Pp. 1151-1211 in *The Collected*

Dialogues of Plato, including the Letters. eds. E. Hamilton and H. Cairns. Princeton: Princeton University Press. 1961.

Sanger, F. "Chemistry of Insulin." *British Medical Bulletin* 16: 183-188. 1960.

White, A., P. Handler, E.L. Smith, and DeW. Stetten, Jr. *Principles of Biochemistry.* New York: MaGraw-Hill. 1954.

b. *Insulin*

Brandenburg, D., and A. Wollmer, eds. *Insulin: Chemistry, Structure and Function of Insulin and Related Hormones.* Berlin: Walter de Gruyter. 1980.

Gueriguian, J.L., ed. *Insulins, Growth Hormone, and Recombinant DNA Technology.* New York: Raven Press. 1981.

Larner, J. "Insulin and Oral Hypoglycemic Drugs; Glucagon." Pp. 1497-1523 in *The Pharmacological Basis of Therapeutics.* 6th edition. eds. A.G. Gilman, L.S. Goodman, and A. Gilman. New York: The MacMillan Co. 1980.

Young, F.G., ed. *British Medical Bulletin* 16: 175-262. (Full issue of volume 16, number 3). 1960.

c. *The Sanger papers*

Ryle, A.P. and F. Sanger. "Disulphide Interchange Reactions." *Biochemical Journal* 60: 535-540. 1955.

Ryle, A.P., F. Sanger, L.F. Smith, and R. Kitai. "The Disulphide Bonds of Insulin." *Biochemical Journal* 60: 541-556. 1955.

Sanger, F. "The Free Amino Groups of Insulin." *Biochemical Journal* 39: 507-515. 1945.

Sanger, F. "The Fractionation of Oxidized Insulin." *Biochemical Journal* 44: 126-128. 1949.

Sanger, F. "The Terminal Peptides of Insulin." *Biochemical Journal* 45: 563-574. 1949.

Sanger, F., and E.O.P. Thompson. "The Amino-Acid Sequence in the Glycyl Chain of Insulin. 1. The Identification of Lower Peptides from Partial Hydrolysates." *Biochemical Journal* 53:353-366. 1953.

Sanger, F. and E.O.P. Thompson. "The Amino-Acid Sequence in the Glycyl Chain of Insulin. 2. The Investigation of

Peptides from Enzymic Hydrolysates. *Biochemical Journal* 53: 366-374. 1953.

Sanger, F. and H. Tuppy. "The Amino-Acid Sequence in the Glycyl Chain of Insulin. 2. The Investigation of Peptides from Enzymic Hydrolysates. *Biochemical Journal* 53: 366-374. 1953.

Sanger, F. and H. Tuppy. "The Amino-Acid Sequence in the Phenylalanine Chain of Insulin. 2. The Investigation of Peptides from Enzymic Hydrolysates. *Biochemical Journal* 49: 481-490. 1951.

Chapter 7

a. *General*

Adelmann, H.B. *Marcello Malpighi and the Evolution of Embryology.* Volumes 1-4. Ithaca, N.Y.: Cornell University Press. 1966.

Adelmann, H.B. *The Embryological Treatises of Hieronymus Fabricius of Aquapendente.* Volumes 1 and 2. Ithaca, N.Y.: Cornell University Press. 1967.

Allbutt, T.C. "Medicine." *The Encyclopedia Britannica.* 11th Edition. 18: 41-64. 1911.

Aristotle. *Generation of Animals.* Cambridge, Mass.: Harvard University Press. 1943.

Aristotle. *Parts of Animals,* trans. W. Ogle. Pp. 41-82 (esp. book 1, chapter 1) in *Aristotle: On Man in the Universe.* ed. L.R. Loomis. New York: W.J. Black. 1943.

Dickinson, E. "I Dwell in Possibility." P. 432 in *The Oxford Book of American Verse.* ed. F.O. Mathiessen. New York: Oxford University Press. 1950.

Eddington, A.S. *The Nature of the Physical World.* New York: The Macmillan Company. 1929.

Frost, R. "Looking for a Sunset Bird in Winter." *Collected Poems of Robert Frost.* New York: Henry Holt and Co. 1929.

Galen. *On The Natural Faculties.* Book 1, Chapter 5. trans. A.J. Brock. London: William Heinemann. 1916.

Gould, S.J. *Ontogeny and Phylogeny.* Cambridge, Mass.: Harvard University Press. 1977.

Harrison, R.G. "Embryology and its Relations. *Science* 85: 369-374. 1937.

Harvey, W. *Disputations Touching the Generation of Animals.* trans. G. Whitteridge. Oxford: Blackwell. 1981.

Hughes, A.F.W. *Aspects of Neural Ontogeny.* London: Logos Press. 1968.

Huxley, J.S. and G.R. De Beer. *The Elements of Experimental Embryology.* Cambridge: Cambridge University Press. 1934.

Katz, M.J. *Templets and the Explanation of Complex Patterns.* Cambridge: Cambridge University Press. 1986.

Katz, M.J. "Is Evolution Random?" In *Development as an Evolutionary Process: M.B.L. Lectures in Biology Series.* eds. R.A. Raff and E.C. Raff. New York: A.R. Liss, pp. 285-315. 1987.

Katz, M.J. and W. Goffman. "Preformation of Ontogenetic Patterns." *Philosophy of Science* 48: 438-453. 1981.

Keele, K.D. *Leonardo da Vinci's Elements of the Science of Man.* New York: Academic Press. 1983.

Kellogg, R. *Analyzing Children's Art.* Palo Alto: Mayfield Publishing Co. 1970.

Mansfeld, J. *The Pseudo-Hippocratic Tract* (Sevens, Ch. 1-11) *and Greek Philosophy.* Assen, The Netherlands: Van Gorcum & Co. 1971.

Mayr, E. *The Growth of Biological Thought: Diversity, Evolution, and Inheritance.* Cambridge, Mass.: Harvard University Press. 1982.

Milne, A.A. *The House at Pooh Corner* (1928). New York: Dutton. 1961.

Morgan, T.H. *Experimental Embryology.* New York: Columbia University Press. 1927.

Needham, J. *A History of Embryology.* 2nd Edition. New York: Abelard-Schuman. 1959.

O'Malley, C.D. and J.B. de C.M. Saunders, eds. and trans. *Leonardo da Vinci on the Human Body.* New York: Henry Schuman. 1952.

Oppenheimer, J.M. *Essays in the History of Embryology and Biology.* Cambridge, Mass.: M.I.T. Press. 1967.

Peck, A.L. *Aristotle: Parts of Animals.* Cambridge, Mass.: Harvard University Press. 1937.

Polya, G. *Mathematics and Plausible Reasoning, Volume I: Induction and Analogy in Mathematics.* Princeton: Princeton University Press. 1954.

Seneca, L.A. *Physical Investigations (Naturales Quaestiones).* Book 3, Chapter 29. trans. T.H. Corcoran. Cambridge, Mass.: Harvard University Press. 1972.

Twitty, V.C. *Of Scientists and Salamanders.* San Francisco: W.H. Freeman. 1966.

Willier, B.H. and J.M. Oppenheimer, eds. *Foundations of Experimental Embryology.* Englewood Cliffs, N.J.: Prentice-Hall. 1964.

b. *Architecture of Limbs*

Ede, D.A., J.R. Hinchliffe, and M. Balls, eds. *Vertebrate Limb and Somite Morphogenesis.* Cambridge: Cambridge University Press. 1977.

Ferguson, A.B., Jr. *Orthopedic Surgery in Infancy and Childhood.* 5th Edition. Baltimore: Williams & Wilkins. 1981.

French, V., P.J. Bryant, and S.V. Bryant. "Pattern Regulation in Epimorphic Fields." *Science* 193. 969-981. 1976.

Harrison, R.G. "Observations on the Living Developing Nerve Fiber." *Anatomical Record* 1: 116-118. 1907.

Harrison, R.G. "Experiments on the Development of the Fore Limb of Amblystoma, a Self-Differentiating Equipotential System." *Journal of Experimental Zoology* 25: 413-461. 1918.

Harrison, R.G. "On Relations of Symmetry in Transplanted Limbs." *Journal of Experimental Zoology* 32: 1-136. 1921.

Harrison, R.G. "The Effect of Reversing the Medio-Lateral or Transverse Axis of the Fore-Limb Bud in the Salamander Embryo (Amblystoma punctatum Linn.). *Wilhelm Roux's Archives für Entwicklungsmechanik* 106: 469-502. 1925.

Harrison, R.G. *Organization and Development of the Embryo.* ed. S. Wilens. New Haven: Yale University Press. 1969.

Nicholas, J.S. "Limb and Girdle." Pp. 429-439 in *Analysis of*

Development. eds. B.H. Willier, P.A. Weiss, and V. Hamburger. Philadelphia: W.B. Saunders. 1955.

Swett, F.H. "Determination of Limb-Axes." *Quarterly Review Biology.* 12: 322-339. 1937.

Chapter 8

a. *General*

Bacq, Z.M. *Chemical Transmission of Nerve Impulses.* Oxford: Pergamon Press. 1975.

Bohm, D. *Causality and Chance in Modern Physics.* Philadelphia: University of Pennsylvania Press. 1971.

Boole, G. *An Investigation of the Laws of Thought on which are founded the Mathematical Theories of Logic and Probabilities* (1854). New York: Dover Publications.

Bridgman, P.W. *The Nature of Physical Theory.* Princeton: Princeton University Press. 1936.

Brooks, C. McC. "Chemical Mediation of the Neural Control of Peripheral Organs and the Humoral Transmission of Mediators." Pp. 151-193 in *Humors, Hormones, and Neurosecretions.* eds. C.McC. Brooks, J.L. Gilbert, H.A. Levery, and D.R. Curtis. New York: State University of New York, University Publishers. 1962.

Cannon, W.B. *The Wisdom of the Body.* Revised and enlarged edition. New York: W.W. Norton. 1939.

Conn, H.J. *The History of Staining.* Geneva, N.Y.: Biological Stain Commission, Humphrey Press. 1933.

Dale, H.H. "Acetylcholine as a Chemical Transmitter Substance of the Effects of Nerve Impulses." The William Henry Welch Lectures, 1937. *Journal of Mt. Sinai Hospital* 401-429. 1938.

Dale, H.H. *Adventures in Physiology, with Excursions into Auto-pharmacology.* London: Pergamon Press. 1953.

Descartes, R. *Treatise of Man.* trans. and commentary by T.S. Hall. Cambridge, Mass.: Harvard University Press. 1972.

Feldberg, W. "The Early History of Synaptic and Neuromuscular Transmission by Acetylcholine: Reminiscences of an Eyewitness." Pp. 65-83 in *The Pursuit of Nature.* eds. A.L. Hodgkin, A.F. Huxley, W. Feldberg, W.A.H.

Rushton, R.A. Gregory, and R.A. McCance. Cambridge: Cambridge University Press. 1977.

Foot, N.C. "Vital Stains." Pp. 564-570 in *McClung's Handbook of Microscopical Techniques.* ed. R.M. Jones. New York: Paul B. Hoeber (Harper and Bros.). 1950.

Fulton, J.F. and L.G. Wilson. *Selected Readings in the History of Physiology.* 2nd Edition. Springfield, Ill.: Charles C. Thomas. 1966.

Hall, M. "The Reflex Function of the Medulla Oblongata and the Medulla Spinalis." *Philosophical Transactions of the Royal Society (Lond.)* 123: 635-665. 1833.

Ham, A.W. and W.R. Harris. "Histological Technique for the Study of Bone and Some Notes on the Staining of Cartilage." Pp. 269-284 in *McClung's Handbook of Microscopical Technique.* ed. R.M. Jones. New York: Paul B. Hoeber (Harper and Bros.). 1950.

Heidel, W.A. *Hippocratic Medicine: Its Spirit and Method.* New York: Columbia University Press. 1941.

Hooke, R. 1665. *Micrographia, or Some Physiological Descriptions of Minute Bodies made by Magnifying Glasses with Observations and Inquiries thereupon* (1665). New York: Dover Publications. 1961.

Hughes, A. *A History of Cytology.* London: Abelard-Schuman. 1959.

Katz, M.J. *Templets and the Explanation of Complex Patterns.* Cambridge: Cambridge University Press. 1986.

Katz, M.J., R.J. Lasek, P. Osdoby, J.R. Whittaker, and A.I. Caplan. "Bolton-Hunter Reagent as a Vital Stain for Developing Systems." *Developmental Biology* 90: 419-429. 1982.

Keele, K.D. *Leonardo da Vinci's Elements of the Science of Man.* New York: Academic Press. 1983.

Keller, A.G. *A Theatre of Machines.* London: Chapman & Hall. 1964.

May, M.T. *Galen: On the Usefulness of the Parts of the Body.* Ithaca, N.Y.: Cornell University Press. 1968.

Mayer, S.E. "Neurohumoral Transmission and the Autonomic Nervous System." Pp. 56-90 (Chapter 4) in *Goodman and Gilman's The Pharmacological Basis of Therapeutics.* 6th Edition. eds. A.G. Gilman, L.S. Goodman, and A. Gilman. New York: The Macmillan Co. 1980.

Mettler, C.C. *History of Medicine,* ed. F.A. Mettler. Philadelphia: Blakiston Co. 1947.

Milne, A.A. "Spring Morning" (1924). *When We Were Very Young.* New York: Yearling/Dell. 1952.

Needham, J. *Man A Machine: In answer to a romantical and unscientific treatise written by Sig. Eugenio Rignano & entitled "Man Not A Machine."* London: Kegan Paul, Trench, Trubner & Co. 1929.

Penfield, W, and L. Roberts. *Speech and Brain-Mechanisms.* Princeton: Princeton University Press. 1959.

Rothschuh, K.E. *History of Physiology.* trans. G.B. Risse. Huntington. N.Y.: R.E. Krieger Publishing Co. 1973.

Schwartz, J.H. "Chapter 10: Chemical Basis of Synaptic Transmission." Pp. 106-131 in *Principles of Neural Science.* eds. E.R. Kandel and J.H. Schwartz. New York: Elsevier. 1981.

Singer, C. *Galen: On Anatomical Preocedures.* London: Oxford University Press. 1956.

b. *The Dale papers*

(Reprinted in: Dale, H.H. *Adventures in Physiology, with Excursions into Autopharmacology.* London: Pergamon Press. 1953.)

Brown, G.L., H.H. Dale, and W. Feldberg. "Reactions of Normal Mammalian Muscles to Acetylcholine and to Eserine." *Journal of Physiology* 87: 394-424. 1936.

Dale, H.H. and H.W. Dudley. "The Presence of Histamine and Acetylcholine in the Spleen of the Ox and the Horse." *Journal of Physiology* 68: 97-123. 1929.

Dale, H.H. and W. Feldberg. "The Chemical Transmitter of Vagus Effects to the Stomach." *Journal of Physiology* 81: 320-334. 1934.

Dale, H.H. and W. Feldberg. "The Chemical Transmission of Secretory Impulses to the Sweat Glands of the Cat." *Journal of Physiology* 82: 121-128. 1934.

Dale, H.H., W. Feldberg, and M. Vogt. "Release of Acetylcholine at Voluntary Motor Nerve Endings." *Journal of Physiology* 82: 353-380. 1936.

Chapter 9

Auden.H. "The Maze." *Collected Shorter Poems, 1927-1957.* New York: Random House. 1966.

Barton, J.K. "Metals and DNA: Molecular Left-Handed Complements." *Science* 233: 727-734. 1986.

Brandenburg, D. and A. Wollmer, eds. *Insulin: Chemistry, Structure and Function of Insulin and Related Hormones.* Berlin: Walter de Gruyter. 1980.

Brown, M.S. and J.L. Goldstein. "A Receptor-Mediated Pathway for Cholesterol Homeostasis." *Science* 232: 34-47. 1986.

Dubos, R. *Man Adapting.* New Haven: Yale University Press. 1965.

Eiseley, L. "The Secret of Life." *The Immense Journey.* New York: Vintage Books. 1959.

Farber, E. *Nobel Prize Winners in Chemistry, 1901-1961.* London: Abelard-Schuman. 1963.

Haeckel, E. *Anthropogenie oder Entwickelungsgeschichte des Menschen.* Leipzig: Verlag von Wilhelm Engelmann. 1910.

Joklik, W.K, H.P. Willett, and D.B. Amos. *Zinsser: Microbiology.* 18th Edition. Norwalk, Conn.: Appleton-Century-Crofts. 1984.

Judson, H.F. *The Eighth Day of Creation.* New York: Simon and Schuster. 1979.

Katz, M.J. *Templets and the Explanation of Complex Patterns.* Cambridge: Cambridge University Press. 1986.

Lehninger, A.L. *Biochemistry.* 2nd Edition. New York: Worth. 1975.

Mayr, E. *The Growth of Biological Thought: Diversity, Evolution, and Inheritance.* Cambridge, Mass.: Harvard University Press. 1982.

Milstein, C. "From Antibody Structure to Immunological Diversification of Immune Response." *Science* 231: 1261-1268. 1986.

Monod, J. *Chance and Necessity.* trans. A. Wainhouse, New York: Vintage Books. 1971.

Nagel, E. *The Structure of Science: Problems in the Logic of*

Scientific Explanation. 2nd Edition. Indianapolis: Hackett Publishing. 1979.

Nobel Foundation. *Nobel Lectures, including Presentation Speeches and Laureates' Biographies. Physiology or Medicine. 1901-1962.* Amsterdam: Elsevier. 1964-1967.

Polanyi, M. "Life's Irreducible Structure." *Science* 160: 1308-1312. 1968.

Sherrington, C. *Man On His Nature.* Cambridge: Cambridge University Press. 1953.

Sourkes, T.L. *Nobel Prize Winners in Medicine and Physiology. 1901-1965.* London: Abelard-Schuman. 1966.

Weyl, H. *Philosophy of Mathematics and Natural Science.* Princeton: Princeton University Press. 1949.

Chapter 10

Cassirer, E. "From a Philosopher" (1942). Pp. 15-18 in E. Lasker, *Chess for Fun & Chess for Blood.* New York: Dover Publications. 1962.

cummings, e.e. "what if a much of a which of a wind." *e.e. cummings: Poems 1923-1954.* New York: Harcourt, Brace & World.

Eiseley, L. "The Secret of Life." *The Immense Journey.* New York: Vintage Books. 1959.

Hobbs, P.V. *Ice Physics.* Oxford: Clarendon Press. 1974.

Hood, T. "I remember, I remember." Pp. 35-36 in *Selected Poems of Thomas Hood.* ed. J. Clubbe. Cambridge, Mass.: Harvard University Press. 1970.

Katz, M.J. *Templets and the Explanation of Complex Patterns.* Cambridge: Cambridge University Press. 1986.

Keats, J. "Letter to John Taylor" (1818). Quoted on p. 538 in *Dictionary of Quotations.* ed. B. Evans. New York: Delacorte. 1968.

Lasker, E. *Chess for Fun & Chess for Blood.* New York: Dover Publications. 1962.

Lord, A.B. *The Singer of Tales.* New York: Atheneum. 1965.

Marshall, R.C., J.W. Faust, Jr., and C.E. Ryan, eds. *Silicon Carbide—1973.* Columbia, S.C.: University of South Carolina Press. 1973.

Poincare, H. *Science and Hypothesis.* New York: Dover Publications. 1952.

Stevenson, R.L. "To any reader" (1885). *A Child's Garden of Verses.* New York: Puffin Books/Penguin. 1952.

Webern, A. *The Path to the New Music.* Bryn Mawr, Pa.: Theodore Presser. 1963.

Wordsworth, W. "Ode. Intimations of Immortality from Recollections of Early Childhood." Pp. 626-633 in *The Oxford Book of English Verse: 1250-1918.* ed. A. Quiller-Couch. New York: Oxford University Press. 1940.

SUBJECT INDEX

QUOTATION INDEX